Fluorescence of Living Plant Cells for Phytomedicine Preparations

Fluorescence of Living Plant Cells for Phytomedicine Preparations

Victoria Vladimirovna Roshchina

CRC Press
Taylor & Francis Group
Boca Raton London New York

CRC Press is an imprint of the
Taylor & Francis Group, an **informa** business

First edition published 2020
by CRC Press
6000 Broken Sound Parkway NW, Suite 300, Boca Raton, FL 33487-2742

and by CRC Press
2 Park Square, Milton Park, Abingdon, Oxon, OX14 4RN

© 2020 Taylor & Francis Group, LLC
CRC Press is an imprint of Taylor & Francis Group, LLC

ISBN: 978-0-367-33171-9 (hbk)
ISBN: 978-0-367-49433-9 (pbk)
ISBN: 978-0-429-31946-4 (ebk)

Typeset in Times
by Deanta Global Publishing Services, Chennai, India

Contents

Acknowledgments

I would like to thank my collaborators in the study of plant cell fluorescence. I am grateful to pioneers in instrumental fluorescence techniques – Drs. Valerii N. Karnaukhov and Valerii A. Yashin. Dr. Karnaukhov organizes the laboratory of microspectral analysis of cells and cellular systems, where all my investigations are carried out. My thanks are to Miss Nadezhda K. Prizova, Mr. Michail V. Goltyaev, Mr. Andrei V. Kuchin and Mr. Nikolai E. Shvirst for their technical cooperation in the experimental work. I owe special gratitude to the Optical Microscopy and Spectrophotometry core facilities, ICB RAS, Federal Research Center, Pushchino Scientific Center for Biological Research of the Russian Academy of Sciences, whose microscopic techniques helped me many times.

Author

Victoria Vladimirovna Roshchina, Doctor of Sciences, Professor, PhD, plant biochemist and physiologist, was born in Leningrad, now Saint Petersburg (Russia). She earned a diploma from Voronezh State University, Russia, as magister and bachelor of Biochemistry and Physiology (1972) in the Biological Faculty. After her postgraduate study in the Institute of Photosynthesis of the USSR Academy of Sciences (Pushchino, Moscow Region, Russia), in 1977 Victoria Roshchina earned her PhD, working on the problem of photosynthetic electron transport. Later, in 1991, she earned her diploma of Doctor of Sciences from the Russian Academy of Sciences Institute of Plant Physiology (Moscow) for the study of the role of neurotransmitters and other natural compounds in chloroplasts.

From 1991 to the present, Dr. Roshchina has actively worked in the Russian Academy of Sciences Institute of Cell Biophysics, Federal Research Center, Pushchino Scientific Center for Biological Research of the Russian Academy of Sciences (Pushchino, Moscow Region) in the field of chemical signaling in plants, including the problems of cell–cell communications with neurotransmitters, autofluorescence of secretory cells, and their reactions to environmental factors such as ozone and other oxidants. She is the author of more than 250 publications, including articles in journals and eight monographs.

In cooperation with Valentina D. Roshchina, she published Russian (Moscow, Nauka, 1989) and English (Berlin, Springer, 1993) versions of the book *The Excretory Function of Higher Plants*, devoted to the modes and mechanisms of the action of the secretory products of plants. In 1991, Dr. Roshchina wrote a Russian version of the book *Biomediators in Plants: Acetylcholine and Biogenic Amines*, published in the Biological Center of the USSR (Pushchino), and then in 2001 an English-language monograph *Neurotransmitters in Plant Life*, published by CRC Press/Science Publishers, Enfield, New Hampshire, USA. Some books have been devoted to ozone action (*Ozone and Plant Cell*, 2003, Kluwer, Dordrecht [coauthored with V.D. Roshchina]) or diagnostics, such as *Cell Diagnostics. Images, Biophysical and Biochemical Processes in Allelopathy* (Eds. V.V. Roshchina and S.S. Narwal), Science Publishers, Enfield, New Hampshire, USA. In 2008, Dr. Roshchina published the first monograph about autofluorescence of plant secretory cells: *Fluorescing World of Plant Secreting Cells*, published by Science Publishers, Enfield, New Hampshire, USA (now a part of Taylor & Francis Group LLC.). In 2014, Springer published her monograph *Model Systems to Study the Excretory Function of Higher Plants*. In 2019, Taylor & Francis Group published the collective monograph *Neurotransmitters in Plants: Perspectives and Applications*, edited by Dr. Roshchina with Dr. A. Ramakrishna.

Introduction

Around the world, as well as modern medicinal practice, interest in phytomedicine, including ancient knowledge, is reviving (Singh 2018; Lazarau and Heinrich 2019). Folk medicine has become part of official medicine in many countries (Hernandez et al. 2012). Studies in this field need to achieve healthcare strategies to gain insight into the public's perception of herbal medicine/the general use of herbs for health, as well as the growing of plants for medicine. As early as 1989, the first monograph was devoted to the problem and reviewed the role of known medicinal plants with phytoncidal properties in the stabilization and restoration of human health in clinics and working spaces (Ivanchenko et al. 1989). The significance of secondary metabolites in medicine today is of interest to many biochemists (Wink 2015). Modern biotechnology has actively used plant molecules from natural tissues for various biomaterials for treatment, such as in wound healing (De Sousa et al. 2019) and general antimicrobial prophylaxis (Bassolé and Juliani 2012; Raut and Karuppayil 2014).

Access to high-quality products should be prioritized. In this case, new approaches to the analysis of phytomaterials are desirable, such as the introduction of a fluorescence method into pharmaceutical practice. The method is based on the luminescence characteristics of natural compounds.

Fluorescence is the emission of light by a substance immediately after the substance has absorbed some radiant energy falling on it. Usually, the short-lived emission disappears just after the actinic light is switched off, unlike phosphorescence, which is a long-lived and stabile process, continuing after the exciting irradiation is stopped. The process of fluorescence is quite different from phosphorescence and bioluminescence. The term *fluorescence* is used when the interval between the act of excitation and the emission of radiation is very small (10^{-8}–10^{-3} seconds).

The modern concept of the fluorescence considers its mechanism to be electron transfer from a singlet-level orbit to a lower-state level. When a compound absorbs the energy of actinic light, some of its electrons are temporarily excited to a higher energy state than their normal state. As the electrons return to their normal energy state, they lose energy, which is emitted by the substance in the form of light emission at a characteristic wavelength. Fluorescence has a longer wavelength than the absorbed radiation that causes it due to the emitted photons being less energetic than the absorbed ones.

Among plant cells, secretory cells contain various fluorescent products, mainly secondary metabolites (Roshchina et al. 1997a, 1998b; Hutzler et al. 1998; Roshchina 2003, 2008, 2012; Talamond et al. 2015). Recently, new aspects of plant autofluorescence have become of interest not only for fundamental investigations but also for practical application, such as express analysis of plant pharmaceutical material (Roshchina et al. 2016, 2017). Plant secretory cells, which contain natural drugs and are easily seen under a luminescent microscope, have been shown to emit all colors of the visible spectrum due to the presence of various fluorescent secondary products (Roshchina et al. 1997a, 1998a; Roshchina 2003, 2008, 2012; Talamond et al. 2015). Similar fluorescent substances are located in various cellular compartments, concentrated in the cell wall, vacuole, and extracellular space. The fluorescence appears to be useful for the identification of compounds in intact secretory cells if there is an appropriate optical system of registration that enables the fluorimetric analysis of the cell *in vivo*.

Fluorescence as a tool for pharmacy may include such parameters as the natural fluorescence of individual components of intact plant material (often termed *auto-fluorescence*) and fluorescence induced after staining with special reagents.

The term *autofluorescence* is used for luminescence of naturally occurring molecules of intact cells in the visible region of the spectrum induced by ultraviolet or violet light (Taylor and Salmon 1989; Haugland 2000). Sometimes, it is determined as the fluorescence of natural compounds within an organism. Tswett (1911) was one of the first to observe autofluorescence of plants under a fluorescence microscope. Today, we have knowledge about autofluorescence in many living cells—in many living fluorescing cells—in microbial, plant, and animal organisms, excited by ultraviolet or violet irradiation and registered by a luminescence technique, especially with luminescence microscopy in various modifications.

Recently, autofluorescence has been presented as a possible tool in the analysis of medicinal plants in pharmacy (Roshchina et al. 2016b, 2017b). Many medicinal plants contain valuable fluorescent pharmaceuticals in secretory cells (Roshchina and Roshchina 1993; Roshchina 2008). Their fluorescence range is wide—from 400 to 700 nm. The cells often show bright fluorescence on excitation by ultraviolet (360–390 nm) or violet (400–430 nm) light (Roshchina et al. 1997a, b, 1998a, 2002, 2007; Roshchina 2003, 2007b, 2008). It occurs due to the presence of secondary metabolites (phenols, alkaloids, terpenoids, etc.). As a possible scheme for differentiation, terpenes, some alkaloids (colchicine according to Croateau and Leblanc 1977), and flavonoids emit in the blue or blue-green region of the visible spectrum, whereas polyacetylenes, isoquinoline, and acridone alkaloids emit in yellow and orange. In the red region of the spectrum, the fluorescence may be due not to only chlorophyll, but also to anthocyanins and azulenes.

If compounds have no color, they are usually not well seen under a conventional microscope without special histochemical staining. In this case, light emission of the cellular compounds may serve as a marker for the cytodiagnostics of the secretory structures under a luminescent microscope, including its novel modifications (Roshchina 2003, 2007a, b; 2008; Roshchina et al. 2007, 2008, 2009). Some of the compounds may also be used as fluorescent probes in studies both for wider laboratory practice and, in particular, for the analysis of the mechanisms of their action on model cells (Roshchina 2005a, b, 2008, 2014).

The analysis of native fluorescence (autofluorescence) has more advantages in comparison with the use of fluorescent proteins from marine organisms as markers (Masuda et al. 2006; Sasaki et al. 2014; Sasaki 2016). In this case, there are no complex and expensive genetic manipulations and little chance of artifacts. It is no coincidence that autofluorescence of specific drugs has been used in thin-layer chromatography to identify them (Wagner and Bladt 1996).

The advantages of the use of autofluorescence as a marker in various studies are as follows. First is the observation of processes without cell damage (plant fluorophores provide researchers with a tool that allows the visualization of some metabolites in plants and cells, complementing and overcoming some of the limitations of the use of fluorescent proteins and dyes to probe plant physiology and biochemistry). Second, there is a fast indication of metabolic changes, because some fluorescing metabolites are affected by environmental factors, so that fluorescence can also be an indicator of stress, damage, etc.). Third, by analyzing images and the emission spectra of natural fluorescent molecules abundant in plant cells, we can gain information on their accumulation and function. Finally, all the above-mentioned advantages may apply to the practice of histology in pharmacology, agriculture, and environmental monitoring.

Another application of fluorescence is used for histological experiments with non-fluorescent or weak emitted cells in plant raw pharmaceutical material. It is the staining with fluorescent reagents for individual cellular organelles or reagents inducing fluorescence of a certain substance.

In this book, we show the practical uses of fluorescence as a tool for living histology and histochemistry in pharmacy. Fluorescent methods may be useful not only for pharmacy as a whole but also for biotechnology and agrotechnology of medicinal plants, which is also important in tropical countries for the education of specialists (Kokate et al. 2011). There are attempts to use fluorescence in the perfume industry and the cosmetics industry (da Silva Baptista and Bastos 2019). The use of the fluorescence phenomenon in pharmaceuticals and cosmetics focuses on toxicology and the use of task-specific fluorescent materials and the development of sensitive and selective analytical and imaging methods. Non-invasive bioimaging using endogenous fluorophores as probes, emerging techniques related to fluorescence, and a brief description of selected fundamental concepts in photoscience are provided for completeness.

Analysis of the fluorescence emitted from secretory cells and secretions of the cells in medicinal plants is based on the comparison of both fluorescent images and the emission spectra in intact plant cells and tissues with those of the individual compounds (drugs) of the species. A researcher can see either the predominant characteristic emission of an individual component (alkaloid, for example) or a complex picture of the emission of many compounds.

Autofluorescence could be used (i) in express microanalysis of the accumulation of secondary metabolites in secretory cells without long biochemical procedures; (ii) in diagnostics of cellular damage; and (iii) in the analysis of cell–cell interactions (Roshchina 2003, 2008; Roshchina et al. 2007, 2008; 2009). The phenomenon is of interest for studying the appearance of the secretory structures and their development in medicinal plants as well as for the accumulation and identification of natural drugs. Fluorescence may also be used for the primary discovery of some biologically active components in complex natural phytopreparations if there are reagents for their identification after staining; for example, histochemical reactions for biogenic amines (Kimura 1968; Kutchan et al. 1986), phenols, lipids, and nucleic acids.

Modern guide-texts (Newal et al. 2002; Duke 2002; Ebadi 2007) and a new book of medicinal plants for pharmacologists (*Microscopic Characterization of Botanical Medicines*, 2011, ed. Upton et al.) do not yet include the fluorescent method for analysis and are based only on microscopic images in transmitted light. An earlier atlas of microphotographs dealing with secretory structures of aromatic and medicinal plants, published by K.P. Svoboda and T.G. Svoboda (2000), also did not include fluorescence, although the study of emission may be of interest to pharmacologists. In recent years, for official and alternative medicine, fast express analysis of pharmaceutically valuable plant species is needed. Up to now, the phenomenon of luminescence emitted from intact plant secretory cells rich in medical drugs and various phytopreparations has not been analyzed in pharmacy. In this monograph, new perspectives of fluorescence application are considered specially for secretory cells of medicinal plants.

The known data and views on unknown aspects of the problem may apply to all aspects of pharmacological and medicinal practice as well as education of students in this and other fields, from botany and biochemistry to ecology. The aim of the book is to show practically the use of fluorescent methods as tools for the histology of plant raw material in pharmacy.

1 Fluorescent Analysis Technique

The visible fluorescence of intact living cells is termed the *autofluorescence* (often induced by ultraviolet [UV] or violet light) of naturally occurring molecules within intact cells or within organisms in the visible region of the spectrum. Recently, this parameter has represented a possible tool in the analysis of medicinal plants in pharmacy (Roshchina et al. 2016b; Roshchina 2017). Many medicinal plants contain valuable pharmaceuticals in secretory cells (Roshchina and Roshchina 1993; Roshchina 2008). Analysis of the fluorescence emitted from secretory cells and secretions of the cells in medicinal plants is based on the comparison of the emission spectra with those of the individual compounds (drugs) of the species. A researcher may see either a prevailing emission characteristic of an individual component (an alkaloid, for example) or a complex picture of the emission of multiple compounds. The aim of this chapter is to show fluorescent techniques as possible tools for pharmacy: either *in vivo* histology (mainly autofluorescence) of secretory cells in medicinal plants or with the application of fluorescent dyes. Autofluorescence could be used (i) in express microanalysis of the accumulation of secondary metabolites in secretory cells without long biochemical procedures; (ii) in diagnosis of cellular damage; and (iii) in analysis of cell–cell interactions (Roshchina 2003, 2008; Roshchina et al. 2007, 2008, 2009). The phenomenon is of interest for studying the appearance of the secretory structures and their development in medicinal plants as well as for the accumulation and identification of natural drugs. In pharmacy, the application of fluorescent dyes or probes for cellular and tissue reactions may also be used in the study of some drugs or important natural compounds (Haugland 2001). In any case, the realization of these directions requires the inclusion of fluorescent techniques in pharmaceutical practice and laboratory investigations.

1.1 FLUORESCENT ANALYSIS TOOLS

The phenomenon of fluorescence may be observed in intact tissues and cells under the usual luminescent microscope or its various modifications. Earlier, registration of the emission spectra was measured in solutions (extracts from tissues) using spectrofluorimeters, but today, there are techniques that enable this to be done *in vivo*. In the following, we shall consider apparatuses that are often applied to the study of fluorescence of living plant systems and extracts from the samples. Among the methods mentioned in the section are those that are most suitable for the practice of pharmaceutical and agrochemical laboratories as well as those that are developing and are rarely used as yet but have potential for the future.

1

1.1.1 APPARATUSES

Luminescence and laser-scanning confocal microscopy (LSCM) may produce images of high quality using fluorescing cells of a luminescence microscope, in which one can see the fluorescing object in a whole field of view (Karnaukhov 1978; Pawley and Centonze 1994; Pawley and Pawley 2006; Roshchina et al. 1972, 1978, 1988, 2001, 2004, 2007; Yeloff and Hunt 2005; Roshchina 2007b, 2008). The principal schemes of the technique are represented in Figure 1.1. Under luminescence microscope, investigators may see and photograph emitted secretory cells among nonsecretory ones and the location of fluorescent secretions in certain compartments within cell. This may be applied to pharmacy in any laboratory. If it is necessary to measure the fluorescence intensity or register the fluorescence spectra, microspectrofluorimetry or some type of confocal microscope will be used. We cannot describe the characteristics of this technique here and refer the reader to specialized

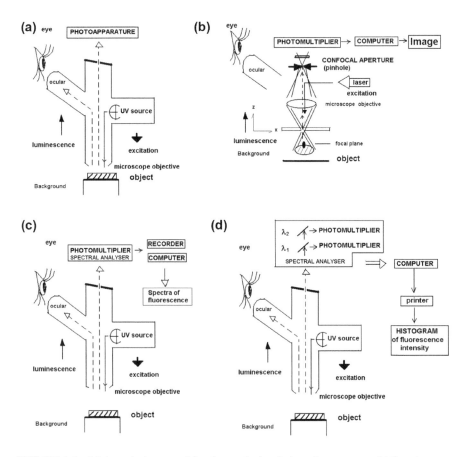

FIGURE 1.1 Main technique used for the analysis of plant fluorescence. (a) Luminescent microscopy; (b) confocal microscopy; (c) microspectrofluorimetry with registration of the spectra; (d) double-beam (dual-wavelength) microspectrofluorimetry with registration of histograms.

literature. Our aim in this chapter is only to show the possibilities of some types of apparatus for pharmacologists.

Table 1.1 shows the types of apparatus used for the study of plant secretory cells. It lists the advantages and difficulties in the use of each technique as well as their industrial potential for future scientific needs. The table allows the reader to evaluate the application of apparatuses for a particular task. As seen from Table 1.1, the problems for determination do not lie only in the construction of the technique. Accessibility and industrial promotion also have a place. We shall demonstrate these methods using concrete examples in Chapters 2 through 4. Looking forward, the attention of scientists also needs to be drawn to new techniques from the perspective of the future.

The first basic method of fluorimetry for living cells was luminescence microscopy. The first steps in the creation of this type of microscopy took place at the beginning of the 20th century. The first observation of chlorophyll fluorescence in plant cells under a microscope was made by the Russian scientist Tswett (1911). Today, he is mainly known as a pioneer of chromatography. The improvement of fluorescence microscopy led to the construction of microspectrofluorimeters for the registration of the emission spectra (Chance and Thorell 1959). The development of microspectrofluorimetry as a method with patents (Karnaukhov et al. 1982, 1983) brought new possibilities for the analysis of intact animal and plant cells, as described in a monograph (Karnaukhov 1978). However, the industrial application range of the apparatuses has been narrow up to now, notwithstanding most of their advantages. This technique is awaiting new employers who are able to make these apparatuses on an industrial scale. Similar non-invasive apparatuses, such as a luminescence microscope in combination with a transmitting microscope, are widespread in practice but cannot provide spectral data, unlike microspectrofluorimetry and spectrofluorimetry. Laser-scanning confocal microscopy apparatuses were developed parallel to the above-mentioned techniques and have been mainly used for fluorescent dyes and probes. Beginning from the year 2000, this technique has been included in many studies on plant systems due to the volume of images of the fluorescing sample structure received by laser excitation in the form of optical slices (then combined as a stack). However, some laser-scanning confocal microscopy apparatuses may record fluorescence spectra.

In the following, we will consider modern developed techniques for intact plant tissue that can be recommended for pharmacy as well as apparatuses that have great potential for the future.

1.1.2 Luminescence Microscopy

Using simple luminescence microscopy, we can see cellular emission (excited with the chosen wavelength) from intact cells with various objectives ×10–×40 and even without a cover glass. In a luminescent microscope, light from the UV source excites the fluorescence of the object. The luminescent image can be perceived directly by the eye of the observer.

The image and color of any fluorescent cell show the changes induced by many experimental factors. Plants enriched in secretory structures with biologically active

TABLE 1.1

Techniques for the Study of Fluorescent Plant Samples Used Today

Apparatus	Aim of the Study	Material for the Study	Advantages of the Methods	Difficulties (Dysfunction) of the Methods
Luminescence microscope (or in combination with transmitting microscope)	Visual and photographic analysis of the emission of intact secretory cells	All parts of plants	Simplicity and non-invasive observation	Absence of spectral data
Laser-scanning confocal microscope	Detailed analysis of the emission of intact and fixed secretory cells with optical slices	Thin lamellar surface of plant objects	Non-invasive observation of the deep layers of studied cells and cellular compartments	Necessity to enclose with cover glass when using ×40 or ×60 or ×100 objective
Microspectrofluorimeter for the spectrum recorder	Spectral analysis of the emission of intact secretory cells	All parts of plants	Non-invasive observation and receiving of spectral quantitative and qualitative characteristics of the sample	Small-scale production of the apparatus
Double-beam (dual-wavelength) microspectrofluorimeter	Comparative analysis of the emission intensity of intact secretory cells in two different wavelengths simultaneously	All parts of plants	Non-invasive observation and receiving of qualitative characteristics of the sample fluorescence	Small-scale production of the apparatus
Spectrofluorimeter for liquid and rigid samples	Spectral analysis of the emission from extracts, the plane surface of plant organs and layers of pollens	Extracts from plant samples or rigid intact and dried tissue	Chemical analysis of drug presence	Disruption of plant samples for analysis or unsuitable thick plane plant organ
Fluorimeter	Analysis of liquid samples at one wavelength	Extracts from plant samples	Dynamics of certain drugs	Narrow wavelength range. Disruption of plant samples for analysis

secondary metabolites have fluorescing products in their cells, and the location of the compounds can be observed.

The fluorescence of objects that lie on the subject glasses (intact tissue or slices) excited by UV (360–380 nm), violet (400–420 nm), or green light (480–520 nm) may be seen in a luminescent microscope with multiplication of objectives ×10, 20, or 40, with water immersion ×85, or with immersion oil ×60 and 85.

Depending on the purpose, slices of fresh plant samples may be observed immediately or after fixation for following analysis. The advantages and disadvantages of the apparatus should be noted. A luminescence microscope (alone or in combination with a transmitting microscope) is used for visual and photographic analysis of the emission of intact fluorescing cells. The technique is simple and non-invasive for the cells, although it is impossible to measure spectral characteristics that may be measured by microspectrofluorimetry and confocal microscopy in some modifications. Fluorescence of intact cells under a luminescence microscope may be excited, for example by UV light, and photographed under a luminescence microscope that allows visualization of the emission after excitation at 340–380 nm. The investigator can distinguish secretory cells among nonsecretory cells and is able to see the object in one plane and photograph the color and location of the fluorescence in certain cellular compartments. Similar non-invasive apparatus (a luminescence microscope in combination with a transmitting microscope) is widespread but cannot be used for receiving spectral data.

1.1.3 MICROSPECTROFLUORIMETRY

The quantitative and qualitative spectral characteristics of living cells may be perceived using a microspectrofluorimeter (Chance and Thorell 1959; Karnaukhov 1972, 1978; Kohen et al. 1974; Karnaukhov et al. 1982, 1987; Weissenböck et al. 1987; Karnaukhova et al. 2010). The apparatus can receive a magnitude fluorescence image of a certain area of the specimen that appears on a spherical mirror. Microspectrofluorimeters, having a detector with optical probes of various diameters up to 2 μm (the changed areas or probe holes composed with the system of mirrors), have been constructed in the Institute of Cell Biophysics at the Russian Academy of Sciences (Karnaukhov et al. 1982, 1983, 1987; Karnaukhova et al. 2010). The emission data may be written in a form of the fluorescence spectra by the help of XY-recorder, which could be coupled with a computer (Figure 1.1). The fluorescence spectra were registered by microspectrofluorimeter MSF-1 at excitation 360–380 nm or 430 nm.

Depending on the aim of the study, the spectral features of the plant cells, secretory structures, and secretions *in vivo* measured with the technique give valuable information about chemical constituents (Roshchina and Melnikova 1995, 1996, 1999; Roshchina et al. 1997; Roshchina 2003, 2007a, 2008). The technique permits the visualization of direct contact of plant cells from foreign species, for example in the modeling of cell–cell contacts (Roshchina 2008; Roshchina et al. 2008, 2009). The fluorescence was excited by UV (360–380 nm) or violet light (420–435 nm). The fluorescence spectra were registered by microspectrofluorimeter MSF-1. The emission intensity (I) in two different spectral regions (in blue-green with maximum

530 nm and in red with maximum 640 nm) of whole microspore cells was recorded on a dual-wavelength microspectrofluorimeter (MSF-2) at room temperature (20–22 C). Intensity in blue-green (or green) and intensity in red were shown as histograms. The intensity of the emission at separate wavelengths may be measured. In this case, statistical analysis was possible for the total of 10–100 pollen grains examined on each slide. Results were expressed as mean ± standard error of the mean (SEM) relative units.

Among the first determinations of flavonoids (*in situ*) by microspectrofluorimetry were the work of Weissenböck and coworkers (1987). Later, this technique was applied to the effects of herbicides on the contents of flavonoids in various weed species of Poaceae (Hjorth et al. 2006).

Recently, Jamme and coworkers (2013) presented deep UV (DUV) autofluorescence microscopy for cell biology and tissue histology, unlike the above-mentioned microspectrofluorimeter, which used the visible spectral range of fluorescence. Synchrotron radiation is a broadband light that can monochromatize in almost any energy range. The DISCO beamline at the SOLEIL synchrotron optimizes the vacuum UV to the visible range of the spectrum, as defined by quartz cutoff from 200 nm to longer wavelengths (600 nm) (Jamme et al. 2010). The energies are accessible on two microscopes at atmospheric pressure. The authors developed a synchrotron-coupled DUV microspectrofluorimeter to which endogenous autofluorescent probes were presented. The probes were distributed in various biological samples, including cultured cells, soft tissues, bone sections, and maize stems. Although most biomolecules do present a contrasting DUV transmission microscopy, few of them will re-emit fluorescence. This autofluorescence is very important in the sense that it permits better discrimination of molecules. Moreover, following autofluorescence makes possible label-free studies of molecules of interest without any external probes or radiolabeling that could impair the activity of the molecule of interest. Although the application of the apparatuses is best known for diagnostic aims in medicine, such as cancer cell visualization, DUV excitation may be used in some cases for measurements of NAD(P)/NAD(P)H or phenols in plant samples: for example, maize sclerenchyma, parenchyma, and phloem studied by the apparatus at an excitation wavelength of 350 nm.

The principal methodical approach to the registration of spectral characteristics is the use of optical probes (seen as a part needed for the analysis) by both microspectrofluorimetry (stationary optical probes up to 2 μm in diameter) and confocal microscopy (regulated size and form of optical probes in one and the same image).

1.1.4 LASER-SCANNING CONFOCAL MICROSCOPY

Unlike the usual luminescent microscope, a confocal microscope has a special confocal aperture (a pinhole), through which the fluorescence of an object excited by laser beam with a certain wavelength passes and is multiplied by a photomultiplier before visualization by the eye. The construction of aperture permits to the light beam be focused on different depth of the object. Unlike microspectrofluorimetry, laser-scanning confocal microscopes may take optical slices of the sample and receive the emission spectra from each slice that is needed for observation. The principal methodical approach to the registration of spectral characteristics is the use of optical probes (seen as a part needed for the analysis) by both microspectrofluorimetry (stationary optical probes up to 2 μm in diameter) and confocal microscopy (regulated size and form of optical probes in one and the same image). The advantages of LSCM as luminescence microscopy derived techniques are as follows: (1) 3-channel simultaneous detection that makes it possible to see images excited by 3 different wavelengths of laser light and to receive a common complicated interference image of the object; (2) the possibility of an increased depth of penetration to receive 20 visual slices (optical sections) or the complete volume (the information must be also quantitatively extracted); (3) interchangeable filters; (4) graphical user interface and the production of accurate computer models as well as mathematical analysis; and (5) pattern analysis of the structure. Earlier apparatuses were able to show images of objects at different depth, but some models also record the fluorescence spectra. Moreover, some confocal microscopes enable the display to be photographed and the image to be analyzed later by the user's choice of optical probes. Biologists usually prefer inverted variants of confocal microscopes for pictures of the living sample.

The fluorescence image of the cells can be seen by air, water, or oil immersion with LSCM. Excitation by three types of lasers: Argon/2 (l 458, 488, 514 nm), HeNe1 (543nm) and HeNe2 (l 633 nm) is often used for the emission. Three photomultipliers were able to detect the fluorescence, either separately or simultaneously by using pseudocolor effects. The image analysis was carried out using different computer programs LSM 510, Lucida Analyse 5 and others. At the excitation by lasers with wavelengths of 405, 458, 488, 543 and 633 nm, the registration of the fluorescence was at range 505–630 nm and 650–750 nm, respectively. Pseudocolours were chosen according to the exciting wavelength blue for 488; green for 543 and red for 633nm. A summed image was seen when the images consisting of the pseudocolours were superimposed and mixed. Main secondary compounds of plants (phenols, terpenoids, alkaloids, etc.) were observed in visible ranges of 400–700 nm with all known techniques using lasers of 360–380 nm or 405 nm.

A well-visualized image of a cellular structure and optical slices of the object allow the demonstration of the changes induced by many experimental factors. Some types of confocal microscopes (Leica, Zeiss, and others) may also register the fluorescence spectra in different parts of the secretory cells. The fluorescence of the cells is observed and measured on subject glasses (slides). All experiments are carried out at room temperature (20–22 °C).

The time of the observation is about 2–4 hours for vital preparation and several days for fixed preparations. All the preparations may be analyzed by three-channel simultaneous detection of a common complicated interference image of the object and details of its structure; by receiving 20 visual slices (optical sections) or the complete volume (the information must be also quantitatively extracted), and by computer modeling the images (Roshchina et al. 2007, 2008).

Another potential method is the use of cryotechniques (Jamme et al. 2013). The application of the method to laser-scanning confocal microscopy has been shown for the study of plant autofluorescence (Corcel et al. 2016; Vidot et al. 2018, 2019). Image analysis of fluorescence emission images acquired between 300 and 650 nm allowed the assignment of fluorescence signals to phenolic compounds based on reference molecules.

LSCMs allow the use optic slices without damage of the objects not more than 1 mm in thickness, especially if one uses microscopes with objectives, that are located below the sample (the inverted variant of microscopes). In the latter type of microscopy, it is also possible to take advantage of special glass cells-cuvettes, usually eight units (each volume about 1 cm^3), into which a user puts the samples, both solid and liquid, and perceives images from lower side of the object. For special studies, it is better to prepare slices mechanically for tissue imaging. For single cells, such as pollen of seed-bearing plants and spores (vegetative, not sexual, unlike male cell of pollen) of spore-bearing species, it is not recommended to enclose them with a cover glass if you want to see their development, and, as well as subject slides, you may use the above-mentioned glass cells-cuvettes. As well as research devoted to particularly to plant pharmaceuticals from land areas, LSCM is used on red algae from seaweed farms near the Caribbean entrance of the Panama Canal (de Vega et al. 2016). The same technique gives information about the fluorescence intensity of the objects at characteristic wavelengths. Some confocal apparatuses allow to receive photo on the display for analyzing the images with optical probes, so-called ROI (ROI = research of interest), on various parts of the sample.

Using confocal and multi-photon microscopy, we may observe invasion of Fusarium fungi in Arabidopsis root vascular system (Czymmek et al. 2007). Laser excitation technique was earlier applied to induce fluorescence of green plants (Chappele et al. 1990).

1.1.5 SPECTROFLUORIMETERS AND FLUORIMETERS

The usual fluorimeters and spectrofluorimeters are suitable for the express-chemical analysis of plant secretions and extracts to test the fluorescent compounds in 1 cm, 0.4 cm or 0.1 cm – glass cuvettes. Alternatively, solutions, as well as plant living or dried samples (0.5–2 cm size of plate) may be analyzed on the subject glasses-slides.

Each sample of plant material (from 2–5 up to 50 mg) extracted at room temperature with solvent (1: 10 w/v) for 10 min (for leaves, petals and sepals of flowers, and roots) or 20–30 min for microspores (vegetative spores of spore-bearing plants and pollen from seed-bearing plants). The extract centrifuged at 1000 g or filtered for further analysis depending on the aim of the study. The absorbance spectra of extracts may be recorded by any spectrophotometer and the fluorescence spectra in visible spectral range – by spectrofluorimeter in 1, 2, 4 and 10 mm cuvettes. The excitation wavelength was 360–380 nm or longer.

Luminescence microscopes show the natural color of fluorescence, unlike confocal microscopes, where pseudocolors are used that only approximately reflect the spectral region of the emission.

1.1.6 FLOW CYTOMETRY

Autofluorescence may be used in the sorting of developed secretory cells in plants by flow cytometry (Bergau et al. 2016). If this is combined with tissue or cell type–specific expression of GFP (green fluorescent protein), the corresponding cells can be isolated by the method of fluorescence activated cell sorting (FACS). In the work of

Bergau and coworkers (2016), the development of glandular trichomes in the tomato genus as a model system was observed using tomato species (*Solanum lycopersicum* and related wild species). The objects have different types of glandular trichomes, of which type VI represent the largest, with a diameter of about 60 μm, and most abundant type. They produce monoterpenes (in *S. lycopersicum*), various sesquiterpenoids, and fatty acid derivatives such as methylketones in *S. habrochaites*. These compounds, particularly in the wild species, contribute to increased resistance against insect pests. In a detailed analysis of the developmental stages of type VI trichomes in *S. lycopersicum* LA4024 and *S. habrochaites* LA1777, differences were observed in the structure of the trichome between the two species. The trichome head is round with a large internal cavity in *S. habrochaites*, whereas it has four clearly visible cells and a small internal cavity in *S. lycopersicum*. The metabolites accumulate in this intercellular internal cavity and contribute to the high metabolite content of *S. habrochaites* and thereby, to better protection against herbivores. In both species, type VI trichomes have a predetermined breaking point between the glandular cells and the intermediate cell that allows the separation of young and mature trichomes. The early stages of type VI development in both species are characterized by the transient presence of a distinct green signal in UV light (Bergau et al. 2015). The identity of the fluorescent compounds could not be determined, but preliminary results suggested they are of flavonoid origin (Bergau et al. 2015). Because these early developmental stages represent a very small fraction of the total trichome population, the green autofluorescence could be used as a specific signal for enrichment of the young trichomes by FACS based on the autofluorescence of trichomes. The procedure included a combination of sieving, density gradient centrifugation, and flow cytometry and allowed us to separate and collect several thousand young and mature trichomes in a short time.

1.1.7 MULTIPHOTON FLUORESCENCE LIFETIME IMAGING (MP-FLIM)

Unlike laser scanning confocal microscopy, which is unable to penetrate into secretory cavity lumina, although the technique has been successfully used to image the outer secretory cells (Goodger et al. 2016), multiphoton excitation has been trialed as an alternative technique due to its potential for penetration to increased depth into plant tissue (Feijó and Moreno 2004). Multiphoton fluorescence lifetime imaging has been described as an excellent tool for imaging deep within plant tissues while providing a means to distinguish between fluorophores with high spatial and temporal resolution (Heskes et al. 2012; Heskes 2014). Multiple fluorescence properties of emitting species can be utilized to provide enhanced image contrast, including emission spectra, fluorescence polarization, and fluorescence lifetime (Levitt et al. 2009). As a whole, fluorescence lifetime imaging microscopy (FLIM) is one of the most informative techniques due to the high sensitivity of a fluorophore's lifetime to its surrounding microenvironment. An extension of this technique is FLIM using multiphoton excitation (MP-FLIM). Multiphoton excitation uses near-infrared wavelengths, which are able to penetrate more deeply into biological tissue than UV wavelengths due to the relative lack of endogenous absorbers and the reduced scattering of near-infrared light compared with UV. Moreover, the nonlinear dependence of multiphoton absorption on the excitation light intensity results in inherent optical sectioning (Denk et al. 1990), often resulting in higher-quality images obtained from deeper within materials, particularly biological tissue, than those obtained through conventional laser scanning confocal microscopy. Reduced photobleaching and increased cell viability are beneficial due to both the limited excitation volume and the less harmful near-infrared wavelengths used in multiphoton excitation (Feijo´ and Moreno 2004). In botany, MP-FLIM has been taken up relatively slowly compared with the other areas of the life sciences; MP-FLIM has mainly been used to detect the close interaction of fluorescently labeled proteins in cells, because there is an increase in signal to noise ratio, which increases the detection limit of the labeled proteins as well as offering improvements in cell viability. Nevertheless, this approach is useful if cells are on the surface of plants and lack strongly absorbing pigments. The full power of multiphoton excitation's increased depth of penetration into plant tissue has not been realized. Today, however, there are examples that have used this technology to obtain detailed lifetime information from fluorophores embedded deep within intact plant tissue. Prime targets for the application of MP-FLIM in plants are the secretory cavity complexes found in many plant families, such as the essential oil cavities of Rutaceae (e.g. *Citrus* spp.) and Myrtaceae (e.g. *Eucalyptus* and *Melaleuca* spp.) and the resin ducts of conifers. Secretory cavities are structurally and metabolically complex, comprising a large extracellular lumen (in which the secretory products, e.g., essential oils, are stored) encased by single or multiple layers of specialized secretory cells (Carr and Carr 1970; Bosabalidis and Tsekos 1982b; Rodrigues et al. 2011). Their embedded nature has traditionally meant that the only way to investigate their structure and function has been through tissue sectioning, resulting in the loss of lumen contents and, particularly in the case of essential oil cavities, disorganization of the bounding secretory cells (Turner et al. 1998). MP-FLIM, with its potential for superior penetration of plant tissue, provides the opportunity to overcome these problems and, moreover, to study the organization of the intact metabolome of cavity lumina. These structures are known to produce and store complex mixtures of secondary metabolites, many of which exhibit autofluorescence. Thus, the use of time-resolved emission detection can potentially enable fine-scale discrimination between different fluorescent species.

As shown by some authors (Heskes et al. 2012; Heskes 2014), ideal candidates for the application of multiphoton fluorescence lifetime imaging to plants are the embedded secretory cavities found in numerous species, because they house complex mixtures of secondary metabolites within extracellular lumina. Although laser-scanning confocal microscopy has been successfully used to image the outer secretory cells (Goodger et al. 2010), it is unable to penetrate into secretory cavity lumina. Therefore, unlike this type of microscopy, multiphoton excitation has been reported as an alternative technique due to its ability to penetrate to increased depth into plant tissue (Feijó & Moreno 2004). Multiphoton fluorescence lifetime imaging has enabled the investigation of the spatial arrangement of metabolites within intact secretory cavities isolated from *Eucalyptus polybractea* R.T. Baker leaves. All species belonging to the *Eucalyptus* genus are of medicinal value. It has been shown that the secretory cavities of this species are abundant (up to 10,000 per leaf), are large (up to 6 nL), and importantly, house volatile essential oil rich in the monoterpene

1,8-cineole together with an immiscible, non-volatile component comprised largely of autofluorescent oleuropeic acid glucose esters. Optical sectioning into the lumina of secretory cavities to a depth of ~80 μm revealed a unique spatial organization of cavity metabolites whereby the non-volatile component forms a layer between the secretory cells lining the lumen and the essential oil. This finding could be indicative of a functional role of the non-volatile component in providing a protective region. Previously, nothing was known about the biosynthesis, transport, or functional role of this component in eucalypt secretory cavities, but the application of MP-FLIM has the potential to aid in characterizing this metabolically complex system. For the first time, the successful application of MP-FLIM with time-resolved emission detection has been successfully applied to visualize the lumen of *Eucalyptus* secretory cavities and used not only to differentiate between the different classes of secondary metabolites housed within lumina but also to show a unique level of spatial organization within these extracellular domains.

Heskes and coauthors (2012) determined the steady-state single-photon excitation and emission spectra for the two components extracted (non-volatile component and essential oil) from secretory cells (secretory cavity lumina) in which fluorescence was excited by different wavelengths, mainly in the 290–300 nm region. The inherent emission from the non-volatile component can be induced by multiphoton excitation, with the maximum emission signal achieved using a wavelength of 760 nm. In order to determine the nonlinear excitation dependence, the emission signal was measured as a function of the excitation light intensity at the sample, indicating that the emission is predominantly induced by two-photon excitation. The localization of the above-mentioned components has provided insight into the inner surface of the lumen, raising questions about its possible function. The use of MP-FLIM has enabled successful imaging of the lumina of intact essential oil secretory cavities for the first time. The increased depth penetration afforded by multiphoton excitation has enabled optical sectioning through isolated secretory cavities to a depth of 80 μm. Autofluorescence occurs from cell wall–bound hydroxycinnamic acids as well as the non-volatile component housed in secretory cavity lumina. A complex of publications is described in the dissertation of Heskes (2014), where it is shown how the non-volatile and volatile components in oil gland cavities are spatially organized in oil glands; this was achieved using the MP-FLIM technique. The method was applied to observe an image of the undisturbed metabolome or oil glands. The distribution of fluorescence in the cavities of *E. polybractea*, and to a lesser extent *E. spathulata*, presented a clear and reproducible pattern. Based on the fluorescence characteristics of extracted non-volatile component and essential oil, the results suggested that the former substance was localized toward the periphery of cavities, while the essential oil is toward the center of the space. In addition, droplets of oil were apparent within the non-volatile component and appeared to be in the process of coalescing with the central domain of oil. It has been hypothesized that this spatial distribution of cavity components could function to minimize the contact between the toxic and volatile essential oil and secretory cells. Besides the non-volatile and essential oil components, the presence and localization of flavonoids in the secretory structures that fluoresced *in vivo* were demonstrated, including the flavonoid pinocembrin. Pinocembrin shows great promise as a pharmaceutical and is predominantly plant-sourced, so *Eucalyptus* could be a potential commercial source of such compounds, as stated by Goodger and coworkers (2016). This non-fluorescent flavanone (Li et al. 2019) was found by the authors in the glands of 11 species of the *Eucalyptus* genus. The foliar glands of this taxonomically distinct group of plants are a rich source of a range of flavonoids and other biologically active compounds with great commercial potential.

1.1.8 COHERENT ANTI-STOKES RAMAN SCATTERING (CARS) MICROSCOPY

Coherent anti-Stokes Raman scattering (CARS) is a nonlinear optical technique delivering a Raman-equivalent signal of distinctively higher intensity, which allows much shorter acquisition times that are suitable for imaging applications, especially for the fast, non-destructive imaging of biological samples. The basis of the method is the Raman spectroscopic technique to determine the vibrational modes of molecules as well as rotational and other low-frequency modes of systems. This is used to provide a structural fingerprint by which molecules can be identified.

The use of Raman spectroscopy in microscopy enables label-free and chemically selective imaging, which has become important in recent years. Molecular identification originates from the specific frequencies of molecular vibrations appearing in a Raman spectrum. A drawback is that spontaneous Raman scattering is a very weak optical effect compared with fluorescence or elastic light scattering. Approximately only one or fewer of 106 photons of scattered light undergoes an energy loss or gain, so-called Stokes and anti-Stokes Raman scattering, through interaction with the molecular vibrations. The imaging of a typical biological sample with Raman microscopy is not always suitable, because it requires very long image acquisition times, even when intense laser beams are used.

In contrast, Coherent anti-Stokes Raman scattering (CARS) apparatus, used for monitoring the differentiation state of cells, generates a coherently driven transition with a nonlinear signal that is in an approximately quadratic relation to the analytic concentration. Usually, it is several magnitudes higher than the corresponding spontaneous Raman signal. The combined action of a pump laser beam at a frequency ωp and a Stokes laser beam at a frequency ωs generates a coherent superposition of the ground state and the first excited vibrational state at the difference frequency $\omega p - \omega s$. Through interaction with the probe beam (here identical to the pump beam), the vibrational coherence is converted into a detectable signal at frequency $2\omega p - \omega s$.

Ebersbach and coworkers (2018) used CARS microscopy as a label-free and non-destructive method to analyze the distribution of secondary metabolites and distinguish between different cell types and chemotypes with high spatial resolution on a single trichome as a secretory structure of industrially cultivated medicinal *Cannabis* plants. Recording the chemical fingerprints of single trichomes offers the possibility to optimize growth conditions as well as to guarantee direct process control. Hyperspectral CARS imaging, in combination with a nonlinear inmixing method, allows the identification and determination of the localization of Δ9-tetrahydrocannabinolic acid in the secretory cavity of drug-type trichomes and cannabidiolic acid/myrcene in the secretory cavity of fiber-type trichomes. Thus, it enables easy discrimination by drug producers between trichomes enriched in the first or the second compounds, respectively. A unique spectral fingerprint has been found in the disk cells of drug-type trichomes, which is most similar to cannabigerolic acid and is not found in fiber-type trichomes. Furthermore, one differentiates between different cell types by a combination of CARS with simultaneously acquired two-photon fluorescence of chlorophyll a from chloroplasts and organic fluorescence mainly arising from cell walls, enabling three-dimensional visualization of the essential oil distribution and cellular structure. Moreover, this method is not limited to *Cannabis*-related issues but can be widely implemented for optimizing and monitoring all kinds of natural or biotechnological production processes with simultaneous spatial and chemical information.

1.1.9 LASER ABLATION TOMOGRAPHY

Luminescent and confocal microscopy generate two- and three-dimensional images of these interactions, but the time required to carry out these measurements is prohibitive for screening a large number of samples. The techniques have limited capability to resolve features below the surface of the sample. Laser ablation tomography is a novel method (Hall et al. 2019) that allows rapid, three-dimensional quantitative and qualitative analysis of tissue. Here, optical coherence tomography (based on the optical backscattering properties of tissue) and a pulsed laser beam that changes in tissue optical properties per second have been used (Boppart et al. 1999).

Laser ablation or photoablation is the process of removing material from a solid (occasionally liquid) surface by irradiating it with a laser beam. Laser ablation can also be used to destroy individual animal cells during the embryogenesis of an organism in order to study the effect of missing cells during development. Experiments have been carried out before, during, and after laser ablation. This method is capable of imaging the anatomy of ~25 root segments per hour and permits the three-dimensional visualization and quantification of root interactions with microorganisms and nematodes (Galindo-Castañeda et al. 2019; Strock et al. 2019). This novel imaging technique provides high-throughput, full-color scans of samples at spatial scales from 0.1 mm to 1 cm with micron-level resolution. The application of the apparatus to studies of root anatomy provides both volumetric quantification and tissue differentiation based on composition-specific autofluorescence.

Autofluorescence from ablating root tissue by UV wavelengths has been recorded in video mode at a rate of 30 frames per second at 1080 pixels with white balance and aperture manually adjusted to optimize the visualization of features of interest. Image stacks have been extracted from these videos, and root anatomical features as well as colonizing organisms have been segmented from image stacks by thresholding based on red, green, and blue spectra with the MIPAR™ software. The spectral values used for thresholding were determined in images by outlining the feature of interest and utilizing maximum and minimum spectral values in the red, green, and blue channels of pixels comprising that feature. Manual review of thresholder images was performed to verify correct segmentation. The original and segmented images have been used for three-dimensional reconstruction and quantification of the root segment and organisms using Avizo 9 Lite software (VSG Inc., Burlington, MA). A new method for high-throughput, three-dimensional phenotyping of roots could enable large screens and genetic studies to characterize host–pathogen interactions. To gain a better understanding of the functional role and genetic control of the interaction between soil organisms and root anatomy, it would be useful to develop high-throughput methods that can spatially quantify and qualify colonization of roots by soil biota in three dimensions.

Examples of the application of the method will also be described in Chapter 2.

1.1.10   POTENTIAL OF SMARTPHONE USE IN THE STUDY OF FLUORESCENCE

A new era of fluorescent instrumentation has begun with the start of smartphone use. The combination of the mobile tool with fluorescence microscopy has great potential. A self-powered temperature-controlled smartphone fluorimeter has been demonstrated by an international group of optic specialists (Hossain et al. 2015, 2016) with leads by Australian scientists Canning and Ast. Miniaturized optical reader devices have been essential for quantifying rapid diagnostic assays and bringing some of the conventional biomedical tests from the bench or laboratory setting to the point-of-care (Wei et al. 2017). A new generation of portable optical imaging and sensing devices has recently been fostered by the surge in smartphone, three-dimensional printing, and rapid prototyping technologies. Mobile phone–enabled devices have been recently developed to diagnose infectious diseases including, for example, HIV/AIDS and malaria as well as for personalized health monitoring.

Such mobile platforms center on rapidly improving the imaging, sensing, and computational power of smartphones as well as the design flexibility and cost-effectiveness of rapid prototyping technologies. These are inexpensive tools that we can recommend now to try in botany and pharmacy. Fluorescence is one of the predominant detection modalities for mobile phone–based diagnostic tools due to the sensitivity and specificity that it enables. Recently, a smartphone intensity fluorimeter has been developed by combining the attributes of a smartphone's inbuilt flash light-emitting diode (LED) as an optical source and the complementary metal-oxide semiconductor (CMOS) camera as a detector. By using the white LED (Jamalipour and Hossain 2019), the instrument not only avoids the requirements of an external source and necessary power supply but also overcomes the issues associated with low irradiance and large mechanical hardware when using other inbuilt sources such as display active-matrix organic light-emitting diode (AMOLED). The broad spectrum of the white LED can filter selectively in order to select any suitable excitation band for different application-specific fluorescence sensors. First, the blue filtered emission ~450 nm is utilized to excite a pH-responsive 4-aminonaphthalimide fluorophore that fluoresces in the green region in the visible spectrum. The pH-responsive green fluorescence at λ ~530 nm may be readily detected using the smartphone camera. Beyond monitoring the water quality at any specific location, the smartphone intensity fluorimeter also demonstrates the capability of generating a real-time pH map by collecting data from multiple points with the global positioning system (GPS) coordinates of the corresponding locations.

Mobile fluorescence imaging devices are frequently used to read DNA or antibody-based immunoassays in various formats such as microarrays, droplets, paper devices, and microfluidic chips (Wei et al. 2017). In addition, mobile fluorescence imaging tools have been used as microscopy platforms to detect nanoscale objects and molecules, including single nanoparticles, viruses, and single DNA strands. The general strategy is to enhance the detection sensitivity of a mobile phone related to a fluorescence imaging device to fewer than 100 molecules per diffraction-limited spot. The principle of the signal enhancement is based on surface-enhanced fluorescence using thin silver films. It permits the preparation of a handheld microscopy device installed on a smartphone, which provides a maximum fluorescence enhancement of approximately tenfold compared with a bare glass substrate. It allows the imaging of individual fluorescent particles. Furthermore, the limit of detection has been estimated to be around 80 fluorophores per diffraction-limited spot by using fluorescently labeled DNA origami structures as microscopy brightness standards.

1.1.11 Materials

Materials for fluorimetry have included subject glasses (glass slides), cover glasses, and glass cells/cuvettes for immersion in glycerin, water, air, or immersion oil. If one needs to see non-fixed motile objects, a researcher can attach similar cells to 2% agar/agars and agar/agarose components that were extracted from the red alga *Ahnfeltia plicata,* differed in their purification degree. If the objects are thick for luminescence microscopy, one could use cover glasses and immersion in the cases of objectives ×20, 40, and 60. The fluorescence of objects on glass slides that are excited by UV (360–380 nm) or violet (400–420 nm) light may be seen in a luminescent microscope with multiplication of objectives ×10, 20, or 40, or with water immersion ×85, or with immersion oil ×60, 85, or 100. If a researcher would like to see the autofluorescence *in vivo* during a certain period, there are several possibilities. One may put the sample onto a glass slide moistened with aqueous medium in a Petri dish with a wet paper filter. The second way is to put it into special eight-camera cuvettes applied to confocal inverted microscopes (cells are put on the glass bottom, and cuvettes are filled with the medium only or 2% agar/agarose moistened with the medium from above). Cultivation techniques of microspores and pollen on glass slides in Petri dishes have been described for confocal microscopy (Roshchina et al. 2004, 2007). Vital images and spectra of fluorescent cells were recorded by laser-scanning confocal microscope on the glass slides at room temperature 20–22°C as published earlier for pollen and vegetative microspores of spore-bearing plant species such as horsetail, *Equisetum arvense* L. (Roshchina et al. 1997, 2004).

1.2 METHODICAL APPROACHES

The main approach to determining the presence of certain molecules in living cells is based on the comparison of the cell emission and the emission of individual compounds alone or in mixtures. Moreover, the autofluorescence of natural secretions depends on the medium of the emitting sample (water, oil, ethanol, etc.). In the following, we will consider some of the methods. Modeling of the secretory compositions and fluorescence in different media is one possible methical approach to the express analysis of pharmaceutical raw materials. The time of plant collection and the degree of maturity of the plant influence the color fluorescence and maxima in the analyzed emission spectra.

It is necessary to keep in mind the difficulties in the assay of compounds based on autofluorescence.

1. First of all is interference of the wavelengths in the range of emission of some phenols and terpenoids fluorescing in blue. This is overcome by extraction of the components with hydrophilic and hydrophobic solvents, as described earlier (Roshchina 2008; Melnikova et al. 1997), and then, measurement of the fluorescence spectra. The disappearance of some of the extracted compounds could provide necessary information.

2. Masking of the weak fluorescence of one of the components in a complex mixture of natural composition by bright emission of another component may take place. In this case, extraction of the masking components using hydrophilic and hydrophobic solvents is also recommended.

3. Differences in the position of maxima in one and the same plant species appear to depend on the media of secretory cells, which may additionally contain oil, phenols, oleophenols, oleoresins, and other compounds found in combination in the cells (Roshchina V.D. and Roshchina V.V. 1989; Roshchina V.V. and Roshchina V.D. 1993). Here, it is helpful to use modeling between the main component analyzed and possible media (Roshchina 2008) to find the correct maximum and shifts. For example, crystals of the phenol rutin emit at a maximum of 620 nm, but in a mixture with menthol oil, they emit at 480–485 nm, while crystals of the alkaloid rutacridone with the main maximum at 595–600 nm had not changed in a similar mixture (Roshchina 2008). Azulene fluoresces with a maximum of 430 nm, but when impregnated into cellulose, it has peaks at 480 and 620 nm (Roshchina 2008).

4. If a researcher needs to know the autofluorescence in different cellular compartments or in excretions, in addition to the optical probes, the isolation of the cell wall and organelles as well as collection of the excretions may be recommended.

5. It appears to be possible to determine artificial or natural mutation with the use of autofluorescence as a tool.

6. One of the non-invasive applications could be the luminescence observation of plant materials in order to know whether or not they are ready for pharmacy, because the main valuable drugs are concentrated in secretory structures. The estimation of cell viability is also a potential use that is especially required for normal pollination and fertilization in genetic analysis for selection in agriculture and biotechnology. Moreover, there is some possibility of using fluorescence changes in the analysis of cell damage, which is necessary for ecological monitoring: earlier diagnostics of the disturbances induced by oxidative stress. Among plant cells, secretory cells contain various fluorescent products, mainly secondary metabolites, but their autofluorescence in intact structures has not been completely described in the literature. Stress, ageing, or pest invasion may influence the fluorescence of secretory products (Roshchina 2008).

1.3 CONCLUSION

The choice of the fluorescent method and apparatus depends on the purposes of a researcher. Today, for pharmacy, the most suitable and most commonly applied methods are luminescence microscopy, microspectrofluorimetry, and confocal microscopy (for intact and living cells and tissues) as well as spectrofluorimetry for big (1–2 cm of plate) plant surfaces and solutions. The author of this monograph has used all these apparatuses in practical work with medicinal plants. Other methods are at the beginning of their use, but have potential. In any case, the autofluorescence or fluorescence of dyes may be successfully studied by the above-mentioned techniques. Multiphoton fluorescence lifetime imaging provides an excellent tool for imaging deep within plant tissues while providing a means of distinguishing between fluorophores with high spatial and temporal resolution. Ideal candidates for the application of multiphoton fluorescence lifetime imaging to plants are the embedded secretory cavities found in numerous species, because they are in complex mixtures of secondary metabolites within extracellular lumina. Previous investigations of this type of structure have been restricted by the use of sectioned material, resulting in the loss of lumen content and often disorganization of the delicate secretory cells. This technique enables determination of whether or not secondary metabolites are partially segregated within these structures.

Of special interest for pharmacy is the simple instrumentation connected with smartphones. In future practice, plant pharmaceuticals will be recognized without lengthy biochemical procedures, even in raw tissue materials or whole organs.

2 Autofluorescence as a Parameter to Study Pharmaceutical Materials

Various groups of substances have significant differences in their spectral features. The observed multi-color fluorescence of intact cells is the sum of emissions of several different groups of compounds, both excreted from or accumulated within the cell and bound to the cellular surface (Buschmann and Lichtenthaler 1998; Buschmann et al. 2000). The biochemistry of components contained in secretory structures has not been well studied yet. However, compounds occurring in the secretions possess characteristic maxima in the fluorescent spectra that are suitable for the identification of the compounds in mixtures, changing the wavelengths of actinic light of luminescent apparatuses for the emission excitation (Figure 2.1). Their fluorescence range is wide.

Cells often show bright fluorescence at the excitation by ultraviolet (UV) (360–390 nm) or violet (400–430 nm) light (Roshchina et al. 1997a, b, 1998a, 2002, 2007; Roshchina 2003, 2007b, 2008). This occurs due to the presence of some secondary metabolites (phenols, alkaloids, terpenoids, etc.). As shown on Figure 2.1, terpenes, some alkaloids (colchicine), and flavonoids emit in the blue or blue-green region of the visible spectrum, whereas polyacetylenes, isoquinoline and acridone alkaloids emit in yellow and orange. In the red spectral region, the fluorescence may be due to anthocyanins and azulenes.

The light emission of cellular compounds may serve as a marker for cytodiagnostics of secretory structures in the luminescent microscope, the microspectrofluorimeter (Roshchina and Melnikova 1995, 1996; Roshchina et al. 1997a, 1998a, b, 2003; Melnikova et al. 1997), or the laser-scanning confocal microscope (Hutzler et al. 1998; Gandía-Herrero et al. 2005a; Roshchina 2007a, b 2008; Roshchina et al. 2007, 2008, 2009; García-Plazaola et al. 2015). It should be noted that they cannot be seen in a conventional microscope without special histochemical staining. Some of the compounds may also be used as fluorescent probes in studies both for wider laboratory practice and specifically, for the analysis of mechanisms of their action on model cells (Roshchina 2005a, b, 2008, 2014). The significance and biological role of plant autofluorescence have been discussed in the literature (Roshchina 2003, 2007, 2008; Iriel and Lagoio 2010b; García-Plazaola et al. 2015; Talamond et al. 2015).

Advantages of the use of autofluorescence as a marker in various studies are (1) observation of processes without cell damage (plant fluorophores provide researchers with a tool that allows the visualization of some metabolites in plants and cells, complementing and overcoming some of the limitations of the use of fluorescent proteins and dyes to probe plant physiology and biochemistry); (2) fast indication

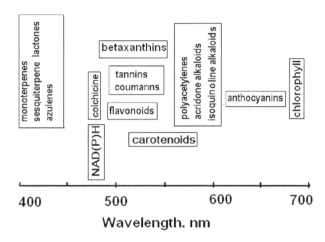

FIGURE 2.1   Position of the fluorescence maxima in the spectra of plant secretory cells. (Modified from Roshchina, V.V., 2008. *Fluorescing World of Plant Secreting Cells*, Science Publishers, Enfield, NH, 2008. With permission.)

of metabolic changes (some fluorescing metabolites undergone by environmental factors, are transformed or new compounds are formed) allowing fluorescence to be also an indicator of stress, damage, etc; and (3) analysis of images and emission spectra of naturally fluorescent molecules abundant in plant cells, providing information about their accumulation and function. Finally, all the above-mentioned advantages may apply to the practice of histology in pharmacology, agriculture, and environmental monitoring. In this chapter, we show the practical use of the phenomenon as a tool for vital histology and histochemistry as applied in pharmacy. Moreover, autofluorescence as a parameter for bioindication has the potential to be used in the food industry (Christensen et al. 2006).

The fluorescence spectra of intact fluorescing cells containing various substances differ in the position of the maxima and the intensity. The influence of the media and various factors on the autofluorescence of secretory and nonsecretory plant cells has been considered in a special monograph (Roshchina 2008). More information may also be gained by modeling natural secretions via the composition of mixtures from various individual compounds (Roshchina 2008, 2014). This permits a preliminary discrimination of the dominating components in the secretory cell tested. The visible fluorescence of a secreting cell depends on the chemotaxonomy and metabolism of the secreted products (Roshchina 2003, 2008, 2014). The first publications related to the identification of phenols in plant cells based on their fluorescence (Weisenböck et al. 1987; Hutzler et al. 1998). Later, the range of compounds analyzed by their emission from living cells became wider (Roshchina 2003, 2008). Fluorescent secondary metabolites occurred in secretory cells may also be applied to vital histological procedures in cell biology. Today, autofluorescence has become an excellent tool for cell imaging and stress diagnostics by, for example, pharmacologists, plant physiologists, and biochemists as well as ecologists.

Depending on the purpose of the study, a researcher can select the method of the study according to the task of the investigation. If it is necessary to observe the fluorescing image as a whole or to find the plant's fluorescing cells among non-fluorescent ones, simple luminescence microscopy and a camera for photos could be used. Looking at the use of autofluorescence in molecular biology, we can determine (1) the prevailing components in plant structures, (2) accumulation of the components, and (3) changes in their component composition after the action of various factors. In recent years, more attention has been paid to secretory cells enriched in fluorescing compounds (Roshchina and Melnikova 1995; Roshchina 2003, 2008; Talamond et al. 2015; Garsia-Peazaola et al. 2015). Up to now, the ability of plant vegetative tissue to fluoresce under UV or violet irradiation in a luminescence micro-scope has been assumed to be associated with chlorophyll. The observation of auto-fluorescence from plant secretory structures opens new possibilities for fluorescent analysis as a source of new discoveries when comparing the autofluorescence max-ima in secretory cells, secretions, and individual compounds found in the chemical composition of secretions peculiar to the plant species and individual tissues. Special attention has been paid to the biological significance of autofluorescence (Iriel and Lagorio 2010a, b) as a signal in a biosystem and a mode for communication.

Meanwhile, secretory cells contain fluorescing substances such as phenols, fla-vones, quinones, and others that make possible fluorimetric analysis of the intact cell *in vivo* (Roshchina 2003, 2008, 2014; Talamond et al. 2015). Figure 2.1 shows the position of fluorescence in the visible spectral region for various groups of fluo-rescent compounds according to their biochemical nature. Cell autofluorescence is specific to any living cell to a different degree, depending on the nature of the fluo-rophores. Thus, in animal tissues, the extracellular matrix often contributes to the autofluorescence emission more than the cellular components, because collagen and elastin have, among the endogenous fluorophores, a relatively high quantum yield. Changes occurring in the cell and tissue state during physiological and/or patho-logical processes result in modifications of the amount and distribution of endog-enous fluorophores and the chemical–physical properties of their microenvironment. Moreover, autofluorescence analysis can be performed in real time, because it does not require any fixing or staining of the specimens. In the past few years, spectro-scopic and imaging techniques have been developed for many different applications both to basic research and diagnostics. One of the difficulties of the analysis is how to identify the emission of individual compounds or certain groups of compounds in complex mixtures due to interference of maxima among components of different chemical nature, but fluorescing in the same spectral region. For example, terpe-noids, NADH, NADPH, and some flavonoids may fluoresce in blue.

Many plant tissues fluoresce due to the natural fluorophores present in cell walls or within the cell protoplast or lumen. While lignin and chlorophyll are well-known fluorophores, other components are less well characterized. Confocal fluorescence microscopy of fresh or fixed vibratome-cut sections of radiate pine needles revealed the presence of suberin, lignin, ferulate, and flavonoids associated with cell walls as well as several different extractable components and chlorophyll within tissues. A comparison of needles in different physiological states demonstrated the loss of chlo-rophyll in both chlorotic and necrotic needles. Necrotic needles showed a dramatic change in the fluorescence of extractives within mesophyll cells from UV-excited weak blue fluorescence to blue-excited strong green fluorescence associated with tissue browning. Comparisons were made among fluorophores in terms of optimal excitation, relative brightness compared with lignin, and the effect of the pH of the mounting medium. Fluorophores in cell walls and extractives in lumina were associated with blue or green emission compared with the red emission of chlorophyll.

To estimate the contribution, one may analyze (1) the emission of extracts (via step by step extractions, in particular by water and by hydrophilic and hydrophobic solvents) in order to identify maxima according to the chemical nature and (2) the emission of secretory cells *in situ* before and after procedures of step by step extraction (see point 1).

2.1   IDENTIFICATION OF SECRETORY CELLS WITH DRUGS AMONG NONSECRETORY ONES

Investigators can distinguish secretory cells among nonsecretory cells and are able to see the object in one plane and photograph the color and location of the fluorescence in certain cellular compartments (Figure 2.2). The common nettle *Urtica dioica* is an ancient medicinal species cultivated for pharmacy (Kavalali ed. 2004). Its stinging hairs (emergences) have acetylcholine and biogenic amines (catecholamines and histamine in the tips of the structures (Roshchina 1991, 2001), but the species also contains formic acid and flavonoids (Muravieva et al. 2007). As seen on Figure 2.2a, under luminescence microscope the base cells of stinging hairs fluoresce brightly in blue-green or yellow when excited by UV and violet light (flavonoids and coumarins emit in this range), while cells near tips and tip cells themselves in hairs – fluoresce in blue. Tip secretion drops within the hair glow especially markedly. Unlike stinging hairs, other types of trichomes (simple retort-like or capitate) in *Urtica dioica* emit weakly and mainly in the dried state. This is a suitable method for pharmacologists to differentiate all hairs in the analysis of both fresh raw and dried material.

Another example of a plant with secretory hairs is the medicinal species *Symphytum officinale* L., also known as common comfrey, Russian comfrey, or woundwort, from the family Boraginaceae. It is known as a medicinal plant, but up to now only the roots have been used as active pharmaceuticals as a wound-healing agent (Golovkin et al. 2001) for the care of disorders of the locomotor system and gastrointestinal tract. The leaves and stems have also been used for the treatment of the same disorders and additionally for the treatment of rheumatism and gout. The impressive wound-healing properties of *Symphytum* are partially due to the presence of allantoin, which promotes cell proliferation; this may possibly be the basis for some of its wound-healing properties. The secretory hairs of the plant are of interest in its structure, because they are surrounded by red-fluorescing hill-like rosette cells with an emission maximum of 680 nm (Roshchina et al. 2011). The color of the fluorescing hairs is green-yellow. The structures luminesce weakly at UV (360–380 nm) and more strongly at violet (420–430 nm) or even green (480–490 nm) light excitation (Roshchina et al. 2011). At violet excitation, the hair emission has maximum of 550 nm and a shoulder of 460–500 nm in the fluorescence spectrum. The structures fluoresce in green at excitation by a 458 nm laser in a confocal microscope (Figure 2.2b,c). Rosettes of base cells with conical hairs are clearly seen on the surface. The

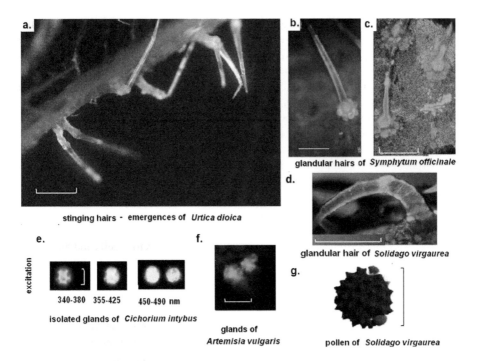

a.

b. c.

glandular hairs of *Symphytum officinale*

d.

stinging hairs - emergences of *Urtica dioica*

glandular hair of *Solidago virgaurea*

e.   f.

excitation

340-380   355-425   450-490 nm

isolated glands of *Cichorium intybus*

g.

glands of
Artemisia vulgaris

pollen of *Solidago virgaurea*

FIGURE 2.2 Images of fluorescent plant secretory cells. (a) View of stinging hairs on the leaf surface of *Urtica dioica* under a luminescence microscope (Leica DM 6000B, bar = 200 μm); (b) and (c) leaf glandular hairs with rosette of basing secretory cells on *Symphytum officinale* under luminescence microscope and sum of images from optical slices under Carl Zeiss LSM 510 NLO confocal microscope (laser 458 nm), respectively, bar = 100 μm; (d) fluorescent trichome of *Solidago virgaurea* under confocal microscope, laser 405, bar = 50 μm; (e) the fluorescence image of multicellular glands on the ligulate flower petal (upper site) of *Cichorium intybus* L. under luminescence microscope at excitation by different wavelengths, bar = 100 μm; (f) sum of images from optical slices of leaf glands of *Artemisia vulgaris* under confocal microscope, laser 405 nm, bar = 100 μm; (g) sum of images from optical slices of pollen grain from *Solidago virgaurea* under Carl Zeiss LSM 510 NLO laser-scanning confocal microscope (laser 458 nm). Different color emission of pollen cover and apertures (from which the pollen tube enters after the beginning of the pollen grain germination) is seen; bar = 100 μm. (From Roshchina, V.V., 2008. *Fluorescing World of Plant Secreting Cells*, Science Publishers, Enfield, NH, 2008. With permission; Roshchina, V.V., unpublished data.)

emission may be related to alkaloids or/and flavonoids of the plant. The secretory cells of *Symphytum officinale* contain alkaloids, among them cyanoglossin, which fluoresces at 460–480 nm (Roshchina and Melnikova 1995), and the flavonoid rutin (Golovkin et al. 2001) may also contribute to the emission. As well as these substances, slime secretions of the plant species contain tannins, which have green fluorescence with maximum at 520–530 nm. All comfrey species contain hepatotoxic pyrrolizidine alkaloids (Betz et al. 1994). These alkaloids, for instance lasiocarpine, as well as chlorogenic and caffeic acids found in *Symphytum* (Golovkin et al. 2001; Tran 2015), may also contribute to the secretion emission. Hair of *Solidago virgaurea* enriched in oil usually fluoresces in blue-green (Figure 2.2d), but oil drops with flavonoids within fluoresce in yellow. The species is rich in flavonols (Wittig and Veit 1999) that show antimicrobial activity (Thiem and Goslinska 2002).

Of especial interest for pharmaceutical analysis are glandular cells. Moreover, sometimes separate glands can be seen under a microscope, as in Figure 2.2e for medicinal species *Cichorium intybus*. The structures have been seen at various excitation wavelengths of the luminescence microscope. On the surface of *Artemisia vulgaris* leaves, researchers can observe the green-fluorescing multicellular glands (Figure 2.2f). Pollen is also a secretory single cell, which releases a pollen tube on fertilization. As seen on Figure 2.2i, under a laser-scanning confocal microscope, the pollen example from *Solidago virgaurea* fluoresces in red at the excitation wavelength of 458 nm, while apertures with the beginnings of pollen tubes fluoresce in green.

There are examples when secretory cells have weak fluorescence in UV or violet light, while the surrounding cells brightly fluoresce. Sometimes, secretions emit in the nonvisible spectral range. Also, secretory cells may look as dark structures on a fluorescent background. In these cases, a researcher must keep in mind the approach of identifying secretory cells using the excitation by light of different spectral ranges. The photos are well seen when dark secretory cells of leaf are surrounded by fluorescent cells, for example for *Mentha piperita* L. (Figure 2.3). By conventional microscopy in transmitted light, glands are seen as dark structures with a yellow interior, while at excitation in blue or green, the secretory cells are easier to observe among the bright surrounding nonsecretory cells.

The emission of secretory rosette cells of gland usually is in blue or yellow. A similar example is also secretory cells of glands on petals in the *Tanacetum vulgare* flower (see Chapter 3, Figure 3.21). When we excited cells with different wavelengths, we saw that weakly fluorescent blue glands may be shown as biseriate structures in light of wavelength 450–490 nm. The surrounding cells are shown better in blue light, perhaps due to the accumulation of terpenoids that emit in blue (Roshchina 2008).

In our experiments, the autofluorescence of glandular structures was analyzed. As shown on Figure 2.3 under transmitted light of useful microscope, glands on

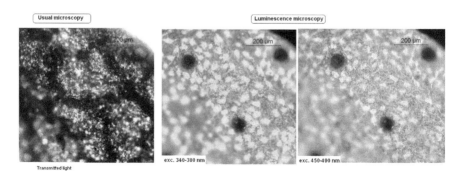

FIGURE 2.3 Fluorescent images of surfaces with secretory cells visualized by luminescence microscopy. *Mentha piperita*—weakly fluorescing glands among surrounding brightly fluorescing nonsecretory cells of leaf, bar = 200 µm.

the leaf of *Mentha piperita* look brownish and rounded. With a large increase in magnification (objective ×40 or 60–100) and glycerin immersion, one can see that the glands consist of many different cells. When excited by UV or blue light, a weak blue fluorescence is observed, while the surrounding leaf surface emits brightly. In three channels of confocal microscope, there is a similar picture of the background.

A similar approach may be applied to studies by laser-scanning confocal microscopy, when a laser of various wavelengths serves as actinic light, although in this case, pseudocolors are used, and the viewer may see rough pictures (non-natural images). The analysis of autofluorescence may be carried out by laser excitation in blue, green, and red. The spectral characteristics of the lasers and channels analyzed depend on the type of technique used. The optical probes chosen for the sample show separate fluorescent parts that are of interest to the researcher.

2.2 COMPOSITION OF FLUORESCING MOLECULES IN THE SAMPLE AND PREVAILING COMPONENTS

The comparison of cell fluorescence with individual molecules is based on the primary knowledge of the presence of individual fluorescing components, even in whole plant species. Histologists, biochemists, and pharmacologists determine the composition in rough plant material by staining for certain chemical groups of compounds (for example, with $FeCl_3$ for phenols or Sudan dyes for lipids) and by extraction with solvents and analysis by chromatography and mass spectroscopy (or nuclear magnetic resonance [NMR] spectroscopy). Special information should be obtained from experiments measuring the fluorescence of individual substances in solution and in solid state. After collection of the data from the above-mentioned procedures, a search for the fluorescing molecule met in the species may be carried out on a single cell, tissue or organ. The composition of the luminescent tissue or single cells may be analyzed by various approaches: (1) changing the excitation wavelengths selecting optical filters (different groups of fluorophores may be activated) and mark color of the emission peculiar to certain compounds; (2) analyzing the emission spectra for the determination of prevailing components in the mixture.

Special experiments dealing with the appearance or missing of some compounds are carried out to observe shifts in metabolism. This occurs during the development of a cell or tissue as well as under the action of environmental factors. Examples of similar experiments will be considered in the following according to the biochemical nature of the substances of the spectrum for the main groups of secondary metabolites occurring in secretory structures as well as those excreted onto the surface of secreting cells (Tunzi et al. 1974; Roshchina 2003, 2008).

The composition of natural compounds in the luminescent tissue or single cells may be analyzed using various excitations of their emission. By changing the optical filters with different wavelengths for emission excitation, a researcher may primarily identify a range of fluorescent drugs, as is seen in Figure 2.4 for the glandular hairs (trichomes) on petals from ligulate flowers of *Calendula officinalis*. Biseriate multicellular hair fluoresces mainly in the blue-green spectral region, although weaker fluorescence is seen in yellow and red. This can be explained by the multicomponent interior of the secretory structure. In blue, phenols, such as tannins and other cell

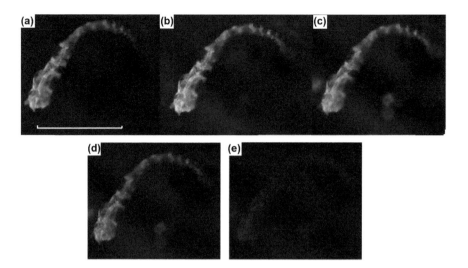

FIGURE 2.4 Fluorescence image of biseriate glandular hair on the ligulate flower petal of *Calendula officinalis* L. (Leica DM 6000B luminescence microscope) at excitation by different wavelengths, nm: (a) 340–380; (b) 355–425; (c) 488; (d) 450–490; (e) 515–560. Bar = 200 μm.

wall polyphenols, weakly fluoresce in blue and green. The intracellular phenolic acid – salicylic acid, found in this plant species (Duke 2002; Murav'eva et al. 2007; Singh et al. 2011), has no detectable luminescence. However, terpenoids and carotenoids, especially α- and δ-cadinenes, as well as flavonoids, coumarins, and quinones, in petals (Singh et al. 2011; Khalid and Teixeira da Silva 2012) potentially emit in blue. Bright green emission may be due to β-carotene (which is abundant in the orange petals [Golovkin et al. 2001] or demonstrates a wider shoulder in the 520–560 nm region [Roshchina 2008]). Although the main orange-red fluorescence is associated with carotene, chlorophyll (which is masked by carotene) may also be identified as a minor component in the petals of *Calendula officinalis* when the sample is excited by light at 515–560 nm. It should be noted that an improved image is seen under a luminescence microscope if the sample is slightly dried in air. Water usually decreases the fluorescence intensity.

The fluorescence intensity (I) in green differs in various plant medicinal species enriched in terpenoids. The green emission may be estimated at wavelength 530 nm within secretory structures and outside of them using dual-wavelength microspectrofluorimetry (Table 2.1). The technique enables this parameter to be compared with chlorophyll emission in red at 680 nm. As seen in Table 2.1, the green fluorescence of terpenoids in oil-containing cells is intensive in green in glandular hair of white petals of flowers from *Achillea millefolium* and weak in rose-colored petals, which also have a significant amount of chlorophyll. A similar difference has been observed for oil reservoirs and secretory hairs of leaves from *Solidago canadensis*. It should be kept in mind that the fluorescence observed in green might also be due to phenols, which are often present in oil-containing cells (Roshchina V.V. and Roshchina V.D. 1993).

TABLE 2.1

Fluorescence Intensity (I, Relative Units) of Terpenoid-Accumulating Cells of Medicinal Plant Species

Plant	Organ	Cell	Common $I_{680/530}$	Glandular Cell I_{530}	Out Glandular Structure I_{530}
Achillea millefolium (Asteraceae)	Flower (petal)	Glandular cell of white petal	2.97 ± 0.16	0.21 ± 0.04	0.06 ± 0.007
		Glandular cell of rose-colored petal	20.3 ± 3.0	0.01 ± 3.0	0
	Leaf	Glandular hair	8.4 ± 0.9	0.26 ± 0.04	0.09 ± 0.03
Solidago canadensis (Asteraceae)	Leaf	Oil reservoir	69.5 ± 3.0	0.26 ± 0.01	0.02 ± 0.013
		Secretory hair	17.4 ± 5.0	0.17 ± 0.06	0

Each medicinal species may contain predominant compounds that are possible to determine based on autofluorescence (Table 2.2). Most of them fluoresce in blue and only a few in green, yellow, and orange-red (Roshchina 2003, 2008). Their emission differs from that of chlorophyll, but is similar to reduced pyridine nucleotides and lipofuscins (pigments of aging), which also fluoresce in the blue and blue-green spectral regions, respectively (Roshchina 2008). In the following, we shall consider the fluorescence of some components *in situ*.

Flavonoids and aromatic acids. Earlier work on plant cellular blue emission related to the flavonoids kaempferol and quercetin was done with luminescence microscopy. Then, derivatives of kaempferol and other phenols, including ferulic acid, were analyzed by laser-scanning confocal microscopy (Hutzler et al. 1998). Recently, new data dealing with the visualization of caffeic and vanillic acids in living samples have appeared (Talamond et al. 2015). Leaf salt glands (peculiar to salt-accumulating plants belonging to the family Chenopodiaceae) fluoresce when excited by light at 360–380 nm. The salt-containing leaf glands of lamb's quarters, *Chenopodium album* L., emit in blue, green, or green-yellow due to the presence of various phenols, including ferulic acid and flavonoids, which impregnate the calcium crystals (which have no emission themselves in pure form) in glandular cells (Roshchina et al. 2011; Roshchina 2014). In flower secretory hairs of *Saintpaulia*, anthocyanin flavonoids predominate throughout the whole multicellular structure. The red flavonoids anthocyanin pelargonidin emits mainly in blue (450–460 nm), and to a small degree in red (625–650 nm). In the base medium of vacuolar sap, the pigment transformed into blue-colored cyanidin shows the same fluorescence characteristics. Aromatic acids such as ferulic, cinnamic, and other are concentrated in the top of the trichome, which emits only in blue under UV light excitation (Roshchina et al. 2017a). Emission attributed to monomeric and/or condensed flavanol is localized in the whole tissue, with major fluorescence in the cuticle region (Vidot et al. 2018, 2019). Image analysis of fluorescence emission images acquired between 300

TABLE 2.2
Fluorescence Maxima of Main Substances Occurring in Plant Secretory Cells (SC) and Excretions (E) of Medicinal Plants

Class of Substances	Representative	λ_{ex}/λ_f (Fluorophore) (nm)	Occurrence		
			In Plant Taxa	In Organs	In Secretory Structures
Alkaloids					
Acridone type	Rutacridone and other acridone alkaloids	381/590–595 (acridone)	Rutaceae (*Ruta graveolens*)	Roots	SC
Isoquinoline type	Berberine, chelerythrine	380/510, 540 (isoquino-linium)	*Berberis vulgaris*	Leaves, roots, stems, and fruits	SC
	Berberine, chelerythrine	380/510, 540 (isoquino-linium)	*Chelidonium majus L.*	Stems and roots	Latex and laticifers
Tropolonic type	Colchicine	360/435	*Crocus autumnalis L.*	Bulbs	SC
Anthocya-nins	Petunidin	Shoulder 510, 555, 570–585, shoulder 610	*Papaver orientale*	Flower	Pollen
Azulenes (Derivatives of sesquiter-pene lactones)	Bee-collected azulenes of pollen	360–380/420, 620,725	*Artemisia, Achillea, Matricaria*	Flowers, leaves, and stems	Oil glands, tri-chomes, pollen
	Chamazulene	360–380/410, 430,725			
	1,3-Chloroazu-lene	360/430			
	Guaiazulene	360/400,433 (lactone)	*Equisetum arvense*	Microspores	Vegetative micro-spores
Carotenoids	Carotenes, xanthophylls	520–560	Many families. *Calendula officinalis*	Flowers	Pollen surface
Coumarins (Dicouma-rins)	Coumaric acid Esculetin 7-Hydroxy-and5–7-hy-droxycoumarins	360/405–427 360/475,(524) 300/479	Many families. *Aesculus hippocasta-num*	Woody plant buds	SC and E
Cytokinins	Kinetin	380/410,430 (adenine)	Many families	All parts of plants	E

(Continued)

TABLE 2.2 (CONTINUED)
Fluorescence Maxima of Main Substances Occurring in Plant Secretory Cells (SC) and Excretions (E) of Medicinal Plants

Class of Substances	Representative	λ_{ex}/λ_f (Fluorophore) (nm)	Occurrence		
			In Plant Taxa	In Organs	In Secretory Structures
Flavonoids	Galangin	365/447–461	Many families	Buds of	SC, ea
	Kaempferol	365/445–450		woody	glands
	Quercetin	365/440,(584)		plants	
Furanocou-marins	4-Methylpsoralen	360/440–420 (furanocou-marin)	*Heracleum sibiricum, Psoralea corylifolia*	Leaves, stems, flowers, and roots	SC, glands, E
C_6-C_3 Hydrocin-namic acids and their esters	Caffeic acid Cinnamic acid Ferulic acid	365/450 360/405–427 350/440 (phenolic ring)	Many families	Buds	E, SC
Indole derivatives	Serotonin	360/410–420 (indole)	Many families. *Urtica dioica* L.	Leaves, stems, flowers, and fruits	E, stinging trichomes
Monoter-penes and their alcohols	Menthol Camphor, camphor derivatives	360–380/415–420 310–313/404–413	Rutaceae. *Mentha piperita.* Asteraceae. Many species. Labiatae/*Salvia*	Flowers, leaves, and stems	SC, glands, E
Polyacety-lenes	Capilline	360/408,430 (two triplet bonds)	*Artemisia capillaris*	Leaves, roots, and flowers	E
Sesquiter-pene lactones	Artemisine	360/395,430 (lactone)	Genus *Artemisia*	Flower	E
Tannins	Valonic acid	360–380/500 (polyphenol)	Many species	Leaves, stems, fruits, and roots	Idioblasts

Sources: Wolfbeis, O.S., *Molecular Luminescence Spectroscopy. Methods and Applications*, Wiley, New York, Chichester, Brisbane, 1985; Roshchina, V.V., *Journal General Biology (Russia)*, 59, 531–554, 1998; Roshchina, V.V., *J. Fluoresc.*, 12, 241–243, 2002; Roshchina, V.V., *Fluorescing World of Plant Secreting Cells*, Science Publishers, Enfield, NH, 2008.

λ_{ex} and λ_f: wavelengths of excitation light and fluorescence, respectively.

and 650 nm allowed the assignment of fluorescence signals to phenolic compounds based on reference molecules. Fluorescent hydroxycinnamic acid is concentrated predominantly in the outer cortex and appears in the cell wall, while fluorescent pigments mostly occur in the epidermis. Apple varieties may be distinguished due to the distribution of flavanols in the sub-cuticle and phenolic acids in the outer cortex.

Anthraquinones. The anthracenic derivatives known mainly as red anthra-quinones are included in the color group of coupled three (frangulin in *Frangula alnus*) or four condensed benzenes (hypericin from *Hypericum perforatum* L., fam. Hypericaceae) (Roshchina 2008, 2014; Roshchina et al. 2016b). On thin-layer plates, the anthraquinone frangulin shows red fluorescence (Wagner and Bladt 1996). Anthraquinones excited by UV light fluoresce in orange-red with maxima 500, 550, and 650 nm (Francis et al. 2014). The autofluorescence of similar compounds will also be considered in Chapter 3, where there are examples of predominant com-ponents in a mixture. The total fluorescence picture in stem particles from the medicinal species buckthorn *Frangula alnus* Mill. (family Rhamnaceae) enriched in the anthraquinone frangulin (Murav'eva et al. 2007) seen under a luminescence microscope shows red- and blue-fluorescent parts (Roshchina et al. 2016b). These pictures can be seen in Chapter 3, Section 3.9, Figures 3.11–3.12. Unlike frangu-lins, another red pigment, hypericin, is concentrated in special glands and reser-voirs of *Hypericum perforatum* L. (family Hypericaceae) in both leaves and flowers (Brockman et al. 1950; Roshchina 2014). The concentrated pigment (see Chapter 3, Section 3.11, Figure 3.14) fluoresces in red with maximum 600 nm.

Terpenoids. Short-wavelength visible emission is also specific to terpenoids—monoterpenes fluoresce beginning from 380–400 nm, sesquiterpenoids have maxima in the 430–460 nm range, and carotenoids fluoresce in green and orange (Roshchina 2008). Exclusively blue fluorescence of cells and secretory cavities occur in Labrador tea, marsh tea, or wild rosemary *Ledum palustre* L., in whose stems the sesquiterpene alcohol ledol (a pharmaceutical substance) predominates in their fluo-rescence under UV light (Roshchina et al. 2017b). The emission of stems has maxi-mum of 430–450 nm, where the main blue fluorescence is due to ledol. At excitation by longer-wavelength light, there is no fluorescence.

Betalains. Brightly colored flowers such as *Bougainvillea*, *Celosia*, *Gomphrena*, and *Portulaca*, which are widely used in folk medicine and as additions to soups and salads around the Mediterranean and in tropical Asian countries, contain the fluorescent pigments betalains (Gandia-Herrero et al. 2005a, b). The chemical group of the compounds includes yellow-emitted betaxanthins and blue-emitted betacyanins, when excited by blue or UV light. Their presence may be connected with successful pollination occurring with the help of insects or birds and is therefore relevant in biocommunication (Iriel and Lagorio 2010a, b). Betalains are the DOPA or dopamine conjugates with phenolic ring. *Celosia argentea* and *Gomphrena globosa* L. (family Amaranthaceae) are used in traditional medicine, including the treatment of respiratory diseases, jaundice, urinary system condi-tions, and kidney problems. *G. globosa* is native to Central America, including regions of Brazil, Panama, and Guatemala, but is now grown globally. Boiled flow-ers of the species form a tea, which is used for baby gripe, oliguria, cough, and diabetes (Roriz et al. 2014). The major betacyanins identified in globe amaranth are gomphrenin, isogomphrenin II, and isogomphrenin III. These compounds are stored in vacuoles in the plant. As a tropical annual plant, *G. globosa* blooms continuously throughout summer and early fall. It is very heat tolerant and drought resistant, but grows best in full sun and regular moisture. *Bougainvillea glabra* has also been used for treatment due to betanidin 6-O-β-glucoside and betanidin (Heuer et al. 1994). The anthocyanins pelargonidin and cyanidin of *Portulaca oleracea*, a succulent species from the family Portulaccaceae, possess a wide spectrum of pharmacological properties, such as neuroprotective, antimicrobial, antidiabetic, antioxidant, anti-inflammatory, antiulcerogenic, and anticancer activities (Zhou et al. 2015). However, there is still little information about the molecular mechanisms of their action.

2.3 INTERFERENCE OF MAXIMA IN SECRETORY CELLS AND MASKING OF THE WEAK FLUORESCENCE OF COMPONENTS

One of the difficulties of the analysis is how to identify the emission of individual compounds or certain groups of compounds in complex mixtures. There is an interference of maxima among components of different chemical nature, but fluorescing in the same spectral region. For example, terpenoids, NADH, NADPH, and some flavonoids may fluoresce in blue.

For the estimation of the contribution, one may analyze (1) the emission of extracts (via step by step extractions, in particular by water and by hydrophilic and hydrophobic solvents) in order to identify maxima according to their chemical nature, and (2) the emission of secretory cells *in situ* before and after these procedures of step by step extraction. The former approach is illustrated by Figure 2.5 for cells of *Mentha piperita* after the use of the solvents water, ethanol, and chloroform. Fluorescence spectra of intact secretory cells have maxima at 460 and 530 nm (Figure 2.5a). The contribution of individual compounds may be understood from the emission spectra of step by step extraction of various solvents. Water extracts have fluorescence at 460–460 nm, specific to terpenoids, flavonoids, and NAD(P)H, while ethanol and chloroform extracts have more maxima. For ethanol extracts, there are maxima at 460, 520, 550, and 680 nm (perhaps flavonoids, flavins, carotenoids, and chlorophyll, respectively) and for chloroform extracts, 560 and 680 nm (carotenoids and chlorophyll).On Figure 2.5b, one can see how the fluorescence intensity of secretory cells of glands changes after testing with the above-mentioned *in situ* extracts. Washing with water leads to approximately a twofold drop in the gland fluorescence at 460 nm and 520 nm, while the following extraction with ethanol and chloroform did not affect the blue-green region, but the chlorophyll emission at 680 nm increased due to absence of masking by water-soluble components. The latter event is related to the liberation from surface components (in particular, some phenols and flavonoids have been extracted). Hydrophilic compounds extracted lipid-soluble secretory products. When hydrophobic solvents are used, mainly chloroform and benzene, terpenoids such as menthol, sesquiterpenes, carotenoids, proazulenes, and azulenes are released during the first 10 min (Roshchina 2008), penetrating into the cell wall and extracellular space. After the chloroform extraction, only structural membrane lipids can fluoresce in the gland, and the fluorescence of chlorophyll decreases to almost nothing.

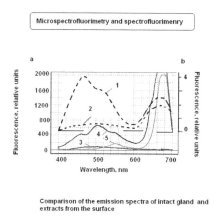

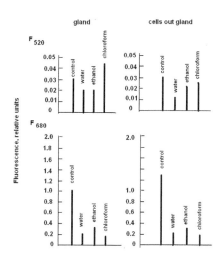

FIGURE 2.5 Comparison of fluorescence spectra of intact cells and extracts of *Mentha piperita* leaf. (a) Microspectrofluorimetry of leaf glands (1) and nonsecretory cells (2); below (b) spectrofluorimetry of leaf extracts (1 g/mL) by water (3), ethanol (4), and chloroform (5). Right – Fluorescence intensity of glands measured by microspectrofluorimetter MSF-2 before (control) and after the leaf extractions by solvents. Extracts excited by light at 380 nm.

In herbal medicine, perfumery, and cookery, mint, *Mentha piperita* from the Lamiaceae family, is widely used, mainly due to menthol giving a characteristic aroma to the essential oil (Murav'eva et al. 2007; Kurilov et al. 2009; Elhoussine et al. 2011). The medicinal properties of tinctures represent natural drugs for spasmolytics in gastrointestinal diseases as well as in complexes of cardiovascular and antiseptic preparations. The herb is also known to be useful for the health when taken in tea form. Thirty compounds were identified (Elhoussine et al. 2011) in leaves including oil (58.61%) and major compounds – terpenoids: menthone (29.01%), menthol (5.58%), menthyl acetate (3.34%), menthofuran (3.01%), 1,8-cineole (2.40%), isomenthone (2.12%), limonene (2.10%), α-pinene (1.56%), germacrene-D (1.50%), β-pinene (1.25%), sabinene (1.13%), and pulegone (1.12%). Turner, Gershenzon, and Croteau (2000) actively studied the accumulation of essential oils in secretory structures of *Mentha piperita* components of different chemical nature, but fluorescing in the same spectral region. For example, terpenoids, NADH, NADPH, and some flavonoids may fluoresce in blue.

Another example is experiments with the medicinal plants *Leonurus cardiaca*, *Tanacetum vulgare*, *Cichorium intybus*, and *Origanum vulgare* (Figure 2.6). First species has glands on leaves and flower. Step by step extractions from the leaves show that the main components were dissolved by chloroform (only smaller components were leached by water and then ethanol) and demonstrated maxima of 450 (terpenoids), 495–500 (flavonoids), and 680 (chlorophyll) nm. In secretory cells of the *Leonurus* glands and in nonsecretory cells, on extraction by water and ethanol the fluorescence was quenched at 480–520 nm, while the emission of chlorophyll increased. Unlike *Leonurus*, water and ethanol extracts from tubular flowers of *Tanacetum vulgare* enriched in water and ethanol-soluble compounds fluoresce with a peak at 460–470 nm. The extraction led to the quenching of the gland fluorescence,

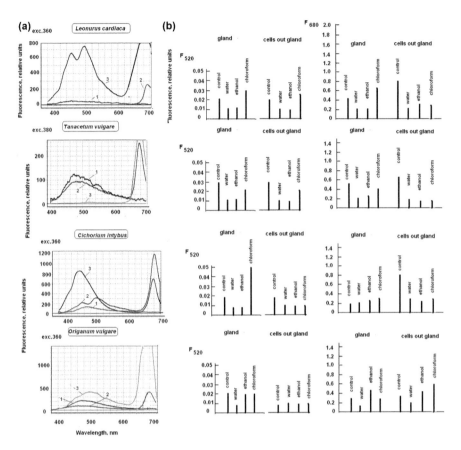

FIGURE 2.6 Fluorescence spectra of extracts (left) and emission intensity *in situ* of glands and nonsecretory cells measured by microspectrofluorimeter MSF-2 (right) as a result of step-by-step extractions by water, ethanol, and chloroform from medicinal plants. Left – Spectrofluorimetry of leaf extracts (1 g/mL) by water (1), ethanol (2), and chloroform (3) Right – Fluorescence intensity of glands before (control) and after the leaf extractions by solvents. Extracts excited by light at 380 nm.

including that of chlorophyll. As for flowers of *Cichorium intybus*, one can see the emission spectra from the water-soluble fraction with the fluorescence maximum at 460–465 nm, the ethanol fraction with maxima 450, 500, and 680 nm, and the highest fluorescence of the chloroform extract with a shoulder at 420 nm (mono-terpenes) and a peak at 450 nm (sesquiterpenes). After extraction with water, the fluorescence of chlorophyll contained in the blue petal surface became visible in the glands, because previously it was masked. The predominant components of the *Origanum vulgare* flowers are more hydrophobic components —extracted by etha-nol, with maxima at 460–470 and 680 nm, or chloroform, with a shoulder at 460 nm (terpenes) and maxima at 500 (flavonoids) and 680 nm (chlorophyll). Moreover, the emission of chlorophyll in the gland is seen even after extraction with chloroform when compared with a control. In this case, we may suppose that all the hydrophilic components are on the surface of glands.

2.4   DIFFICULTIES IN COMPARISON OF THE EMISSIONS OF INTACT SECRETORY CELLS, SECRETIONS, AND INDIVIDUAL COMPONENTS: MODELING WITH MIXTURES

In many cases, intact secretory cells show multicomponent fluorescence spectra, and it is necessary to know the chemical composition of the studied plant species and the predominant components in order to compare them with the emission of individual components (Roshchina 2008). In this process, there are many limitations. We cannot compare quantum yields of the compounds met in secretory cells, because it is impossible to determine them. Data about the absorption spectra of the intact cells and the secretions are absent. Moreover, the modern concept of quantum yield applies only to solutions of the pure compounds studied. Also, the fluorescence behavior in the mixtures of the compounds is not yet clear.

Compounds occurring in the secretions possess characteristic maxima in the absorbance and fluorescence spectra, which is favorable for the identification of the compounds in mixtures. The analysis of the mixtures of the secretory components in natural secretions may be based on the knowledge of components that can quench or stimulate fluorescence. The fluorescence of the components of secretions may change during the development of the cell, in particular, and of the plant organism as a whole (Roshchina 2008).

In intact secretory cells, the fluorescence is dependent on the environment, and the fluorescence of individual compounds in the secretions depends on the medium. The chemical nature of the medium (water, oil, ethanol, and methanol, etc.) and the composition of the secretion can define the character of the fluorescence. An analysis of the effects of the medium has been done in a monograph (Roshchina 2008). If the secretion represents oils (non-fluorescent), including phenols such as the flavonoid rutin or the sesquiterpene lactone azulene or the alkaloid rutacridone, the emission of the oil medium differs from the ethanol extracts. We can see the shift of the fluorescence maxima to shorter wavelengths. For example, in the medicinal plant *Ruta graveolens*, crystals of rutin fluorescing in orange with maximum at 595–600 nm. When it interacts with menthol, the oil of around the rutin crystal becomes blue fluorescent, with maximum at 455–460 nm (Roshchina 2008). The influence of cell wall fluorescence has been estimated by the light emission of hemicellulose and pectin as the main components (pure cellulose fluoresces in blue at 430–460 nm, depending on the level of purification). Crystals of the components fluoresce in blue at 408 and 450–460 nm at excitation by light at 360–380 nm. In a water emulsion, hemicellulose and pectin have a weak blue fluorescence, because water quenches the emission.

Azulenes are usually present in the essential oils of secretions (Roshchina V.V. and Roshchina V.D., 1993). The addition of silicon oil to their crystals leads to shifts in fluorescence to UV emission instead of blue (Roshchina 2008). Often, the cell walls of secretory cells are also saturated with oils, which may include azulene. In this case, the fluorescence in the variant crystal+oil+pectin shifts to 490 nm. When hemicellulose is added to this mixture, the whole composition fluoresces in a shorter blue spectral region at 410–420 nm. If the azulene is impregnated into pure cellulose (paper filter, for example), maxima in the blue-green and red spectral regions appear (Roshchina et al. 1995). Experiments with the alkaloid rutacridone showed that water and ethanol quenched the orange fluorescence of the crystal, while oil induced a weak decrease, but the surrounding oil solution of the alkaloid had a new maximum 460 nm (Roshchina 2008). A similar picture was seen for the fluorescent spectra of rutacridone in ethanol solution, although maximum 460 nm was converted to a shoulder.

In other cases, the oxidation factor also influences, in particular, when reactive oxygen species arise on the cell surface. For example, azulene in various solutions undergoes the addition of hydrogen peroxide (Roshchina and Roshchina 2003; Roshchina 2008). Water quenches the fluorescence of the sesquiterpene lactone, whereas a water solution of the peroxide stimulates it.

It is also possible to estimate the contribution of membrane fluorescence. London and Feigenson (1981) demonstrated that nitroxide spin-labeled phosphatidylcholine, a usual membrane component that also forms aqueous lipid vesicles *in vitro*, quenches fluorescence at 500 nm and shorter wavelengths (excitation by light of 340, 358, 430, and 335/386 nm) in models of lipid membranes. The quenching primarily arises from the nearest spin-labeled lipid neighbors to the fluorophore.

2.5 ACCUMULATION AND DRYING SECRETIONS IN SECRETORY CELLS

Most medicinal plants, where pharmacologically valuable secondary metabolites are accumulated, have different autofluorescence depending on the stage of development (Roshchina et al. 1998a, 2011; Roshchina 2008). The accumulation of different fluorescing components may be seen from the fluorescence spectra. The various spectra of fluorescence from glands in different species have been observed to depend on the stage of maturity. This has been seen in examples of terpenoid-enriched *Heracleum sibiricum* and *Calendula officinalis*, in phenol-containing scales of buds from *Betula verrucosa*, cones of *Humulus lupulus* L., and herb *Echinops globifer*, and in resin-containing secretory cells of *Pinus sylvestris* (Roshchina 2008).

Various emissions have been observed after the appearance of formed secretory cells on the petals of flowers from the medicinal plant *Achillea millefolium* (Roshchina 2003). The cells fluoresce blue-green, and when developing, they also emit in the red spectral region. Azulenes are absent in immature glands and appear in developed secretory structures. In oil, they may fluoresce in blue at 420–430 nm, but when impregnated into the cell wall, they fluoresce in red (Roshchina et al. 1995). Red fluorescence of the glandular cells arises only in mature glands. There is accumulation of not only proazulenes and azulenes, but also anthocyanins in fresh pharmacological materials from *A. millefolium*. In this species, flowers may have rose-colored petals that contain blue-green fluorescing glandular cells full of components that differ from those of white petals (Roshchina 2008). Blue-green fluorescing glands of white petals emitted 20-fold more intensively. In future, pharmacologists may prepare special scales of fluorescence for calibrating azulenes and carotenoids, whose concentrations change during maturation of the secretory cells (Roshchina et al. 1998a).

The observed fluorescence of intact cells is a sum of emissions of several different groups of substances, both excreted from or accumulated within the cell and bound to the cellular surface. The visible emission from secreting cells may occur from the secretory products themselves located on the cell surface, in extracellular space, or within secretory vesicles and vacuoles. Some contribution to the process is also due to structural components of the cells—cell wall, plasmalemma, and cytoplasm with some organelles.

The fluorescence spectra of the secretory compounds can differ strongly depending on the emitting fluorophore. Most plant secondary substances can fluoresce in the visible region of the spectrum. The lack of fluorescence is due to certain causes (Wolfbeis 1985). The first is a lack of electron systems, as in the case of fatty alcohols, some steroids, and pheromones. The second cause may be the presence of rather short and non-fluorescent π-electron systems, as in the case of fatty acids; aldoses and others; the mono-, di-, and some trienes; vitamin C; and most of the amino acids. In addition, the presence of a transition metal ion in a fluorescent system may render it non-fluorescent (e.g. heme and related transition metal complexes). All the rules for fluorescence follow. Substances can emit in all states if they have discrete energetic spectra; otherwise, the energy is transformed into heat energy instead of fluorescence. According to Pheophilov (1944), many substances, in particular dyes, fluoresce in rigid solutions (sugar, gelatin, plastic, etc.) and in the absorbed state. The possibility of dyes fluorescing in such states has been observed if their structure permits rotation of two large parts of the molecule surrounding the chain of conjugated double bonds, which determines the color of the compound. The light emission depends on the chemical groups of the compound. Aromatic hydrocarbons fluoresce if they include amino, alkyl, and oxy groups (Goryunova 1952). However, the inclusion of nitro, nitroso, and halogen groups may quench the fluorescence (Zelinsky 1947). The aging pigments, which are formed by the reaction of malondialdehyde with free amino acids, in particular in plant tissues fumigated by ozone, fluoresce in blue (Roshchina and Roshchina 2003). In many plant excretions, there are non-fluorescent polysaccharides, which are usual in slimes or mucilages. The slimes also include proteins, which in many cases are non-fluorescent too.

The autofluorescence of fresh and dried materials differs because, except for oil secretions, water usually quenches the emission to some degree. Depending on the species, a researcher can observe more information about fluorescent images in dried material.

Phenols. Phenols are compounds that have one or more phenolic groups (benzene rings with hydroxyl groups). This class of substances is subdivided into several groups: aromatic acids, flavonoids, coumarins, furocoumarins or furanocoumarins, and tannins and their derivatives. The most important single group of phenols in plants is the flavonoids, which consist mainly of catechins, proanthocyanidins, anthocyanidins, flavones, flavonols and their glycosides. For instance glycosides of anthocyanidins are known as anthocyanins. Among the variety of naturally occurring oxygen ring compounds, coumarins, furocoumarins, and flavonoids fluoresce intensely (Wolfbeis 1985). Chromenes and chromones also have native fluorescence. 2,2-Dimethyl-2H-chromene is non-fluorescent, but the introduction of an 8-methoxy group gives rise to fluorescence (Wolfbeis 1985). Flavonoids fluoresce in blue, but some of them also have maxima in green and yellow-orange (see following). The bioavailability, metabolism, and biological activity of flavonoids depend upon the configuration, total number of hydroxyl groups, and substitution of functional groups about their nuclear structure (Kumar and Pandey 2013). Most recent research has focused on the health aspects of flavonoids for humans. Many flavonoids have anti-oxidative activity, free radical scavenging capacity, coronary heart disease prevention, hepatoprotective, anti-inflammatory, and anticancer activities, while some flavonoids exhibit potential antiviral activities. In plant systems, flavonoids help in combating oxidative stress and act as growth regulators. For pharmaceutical purposes, the cost-effective bulk production of different types of flavonoids has been made possible with the help of microbial biotechnology. For more complete observation of phenolic compounds met in plant secretions, see the special monograph (Roshchina 2008).

Monoterpenes. Volatile aroma terpenoids, mainly monoterpenes such as citral, geranial, linalool, geraniol, geranial, nerol, etc., are found in leaves and flowers (Harborne 1993; Roshchina V.V. and Roshchina V.D. 1993). Terpenoids such as the monoterpenes geranyl acetate, neral, geranial, geraniol, and their acetates are found in pollen grains of various families (Dobson et al. 1996). Flower scent and liquid excretions enriched in various terpenoids include monoterpenes such as linalool and its derivatives (Fischer 1991; Knudsen et al. 1993; Borg-Karlson et al. 1993). According to (Fischer et al. 1989; Fischer 1991), volatile aroma terpenoids also found in flowers, such as monoterpenes, can be depressants of plant growth and germination in millimolar concentrations, whereas sesquiterpene lactones showed both stimulation and inhibition of growth processes, depending on concentration. Also, flower volatiles contain dimethyl disulfide and trisulfide, C_{10}–C_{13} alkanes (Stransky and Valterova 1999).

Most of the monoterpenes, especially nerol, linalool, and geraniol, are concentrated in flower essential oils (more than 20%), whereas geranial and geranyl acetate are concentrated in pollen (Roshchina V.V. and Roshchina V.D. 1993). Some of them, for instance linalool and geraniol, are attractants for insects. Monoterpenes occur in essential oils of many secretions from all parts of species belonging to the Asteraceae. Cymol is a component of the essential oils of the blossoms of *Achillea millefolium* (Harborne 1993) and *Artemisia ludovicjana*, up to 33–50% of total essential oil (Elakovich 1988). Camphor is also a major component of the essential oil of *Artemisia maritima* (Murav'eva et al. 2007).

The mechanism of the action of monoterpenes appears to be in the interaction with chromosomes, because abnormalities in chromosomes have been observed (Elakovich 1988). The volatile aroma of many flowers containing monoterpenes, for instance cineol, depressed cholinesterase activity. Monoterpenes (geraniol, nerol, and linalool) can regulate the coupling of electron transport and phosphorylation in chloroplasts and mitochondria (Roshchina V.V. and Roshchina V.D. 1993).

Sesquiterpene lactones. Sesquiterpene lactones demonstrate high biological activity (Rodriguez et al. 1976; Elakovich 1988; Duke et al. 1988; Fisher et al. 1989; Fisher 1990, 1991). Many of them show pesticidic features, insect repellency, and antimicrobial and antihelminthic properties (Duke et al. 1988). Depending on the concentration, pure sesquiterpene lactones may regulate seed germination (Fisher 1990).

Sesquiterpene lactones at 10^{-9}–10^{-5} M stimulated the seed germination of obligate root parasites (Fisher et al. 1989). Major mechanisms of their action consist of interaction with nucleic acid and influence on protein synthesis (Roshchina and Roshchina 1993). Some of the sesquiterpene lactones are proazulenes, and their derivatives are the azulene pigments (Rybalko 1978; Konovalov 1995). Colorless proazulenes may be converted into blue azulenes. They are condensed structures, consisting of five and seven carbocycles and their derivatives. Colors of azulenes, from blue and violet-red, are related to the maxima at range 550–610mm in their absorbance spectra. Azulenes are present in essential oils of glands in some Asteraceae species (Konovalov 1995), in needles of conifer plants (Kolesnikova et al. 1980; Roshchina 1999), and in pollen grains of many plants (Roshchina et al. 1995). The blue color of conifer needles, earlier supposed to be due to light reflectance by surface waxes (Reicosky and Hanover 1978), may be related to azulenes. The substances are abundant in flowers, especially in glands and glandular structures of petals. Pollen also contains proazulenes (Konovalov 1995) and azulenes (Roshchina et al. 1995; Roshchina 1999). The localization of azulenes varies from oil cells (including idioblasts) and schizogenous oil ducts (Reichling and Beiderbeck 1991) to the surface of pollen (Roshchina et al. 1995) and even chloroplasts of some plants (Roshchina 2001).

2.6 MEDIUM AND FLUORESCENCE

The possible contribution of medium to the fluorescence of secreting cells has not been studied yet. The compounds in the secretions possess characteristic maxima in the absorbance and fluorescence spectra, and medium also may influence on the emission of individual components (Roshchina 2008).

Visible changes in the spectra may include the effects of quenching or stimulation of the fluorescence of organic compounds (Roshchina 2008). Quenching occurs at the co-striking with other molecules, for instance with oxygen or water. In a result of the event, it is the fast escape of electrons on the surface at the contacts. Moreover, the state or phase of the compound plays a significant role in the quenching of the fluorescence. If the substance is in a solid state, for instance, some flavonoids, such as rutin and quercetin, have maximum emission in the yellow-orange spectral region near 580 nm, but other flavonoids do not. Moreover, certain flavonoids do not demonstrate any fluorescence, because as a result of internal conversion, a non-emitted transition of energy takes place. More rigid, well-organized systems emit to a lesser degree than disorganized systems (especially solutions). In any case, an increase in the possibility of non-emitted transitions leads to a quenching of luminescence. This possibility depends on increased temperature (temperature quenching), concentration of luminescent molecules (concentration quenching), or admixtures (admixture quenching). Among internal factors of quenching is the inclusion of quenching groups in aromatic compounds (Zelinskii, 1947). For example, similar inclusion of NO group (with large dipole moment) completely diminishes the fluorescence, whereas the inclusion of carbonyl- or carboxyl- groups only decreases the emission. Among the substances that can quench fluorescence (Lakowicz 2006) are water, oxygen, and others. The quenching of the fluorescence by water is explained by the fast escape of electrons from the substance located on the contact surface with water molecules. Oxygen also can quench light emission. Singlet oxygen and hydrogen atoms in C-H and O-H bonds may quench fluorescence (Ermolaev et al. 1977). Various metabolites (histidine, cysteine, fumarate, etc.) can also quench fluorescence, and many groups of chemicals (indole, carbazole, and their derivatives) are sensitive to quenchers (Lakowicz 2006). The quenching appears to occur by the donation of an electron from the fluorescing substance to a quencher. Oxygen is a significant quencher of luminescence (Kautsky 1939) as well as hydrogen peroxide, especially studied for proteins and indoles (Eftink and Ghiron 1981). The fluorescence of flavin adenine dinucleotide (FAD) and reduced nicotinamide adenine dinucleotide (NADH) may be quenched by the adenine moiety (Lakowicz 2006). Some ions also may quench fluorescence (Eftink and Ghiron 1981; Eftink 1991). Organic radicals may quench the electron-excited state of many fluorescent dyes (Buchachenko et al. 1967). In particular, donors of electrons, such as anthracene, fluorescein, and rhodamine, fluoresce, but their emission is quenched by acceptors of electrons; for instance, radicals of nitric acid (nitrate radicals).

The quenching of fluorescence in a solution may be enhanced by increased temperature, admixtures, change in the pH of the medium, and increased concentration of fluorescing compounds (so-called internal concentration quenching). The mechanisms of the phenomenon for water may be in the polarity of the chromophore environment and transfer between the excited chromophore and water (Dobretsov et al. 2014). The quenching appears to be a result of chromophore–solvent hydrogen bond breaking in the excited state.

The fluorescence of organic compounds occurs depending on the structure of the molecule and is specific to rigid structures that exclude the possibility of nonradiated transfer. For example, more free rotation of part of the molecule induced the quenching of fluorescence. The maxima of the flavonoid fluorescence shifted from blue to the orange spectral region.

An increase in the fluorescence of secretory components in mixtures has also been observed, but as yet has not been specifically studied. For example, anthocyanins stimulate fluorescence in phenolic mixtures. The substances, which have no free hydroxyl groups in the C15 position, can fluoresce in a water–citric acid phase with peaks 510, 515, 555, at range 570–585 nm and with shoulder 610 nm. Anthocyanins in contact with metals form complexes that stimulate light emission. Flavones interact with anthocyanins, leading to stimulation of the anthocyanin luminescence. However, the interaction of anthocyanins with quercetin quenches the anthocyanin fluorescence. Possible causes of the enhanced fluorescence in the mixtures of the secretory products *in vivo* are the electron transfer from one compound to another and the formation of complexes.

Some natural compounds fluoresce to different degrees depending on the solvent, crystal, or liquid state. Certain experiments based on modeling in mixtures of individual substances have demonstrated here. Examples are fluorescence spectra of the flavonoid rutin and the sesquiterpene lactone azulene (Roshchina 2008). One could see shifts in maxima of the spectrum or the appearance of new maxima. Crystals of rutin emitted in red, while ethanol solutions of the compounds showed two maxima in blue (460 nm) and orange (550 nm). Oil (menthol or silicon) shifted the first maximum of the crystal in blue to 440–450 nm and the second peak to 620 nm. Crystals of synthetic azulene emitted with maximum of 420–430 nm and in ethanolic solutions may have a maximum at an even shorter wavelength. Synthetic azulene and azulenes isolated from bee pollen loads (pollen collected by bees as a food) were impregnated into cellulose paper filters and studied by microspectro-fluorimetry at an excitation of 360–380 nm (Roshchina et al. 1995). In many cases, azulenes from pollen loads fluoresced in blue like artificial azulene. However, when they were impregnated into cellulose as a model of the cell wall, an additional maxi-mum of 610–620 nm (in red) was also seen. In non-fluorescent silicone oil, artificial azulene fluoresced in UV, with maxima shorter than 400 nm, although crystal weak emitted in the region 410–430 nm. Pectin added to crystal of azulene also may shift the pigment maximum to a longer spectral region (in this case – 480 nm).

2.7 FACTORS THAT ACT ON AUTOFLUORESCENCE OF NATIVE CELLS

Autofluorescence of plant pharmaceutical material depends on both internal (growth and development) and external factors (Roshchina 2008). In the following, their effects will be considered.

2.7.1 INTERNAL FACTORS

Metabolism varies during phases of plant development, and new compounds appear, showing alterations in maxima positions in the fluorescence spectra of intact tissues and cells (Roshchina 2008). This may be used for express analysis of the maturity of the plant material. On Figure 2.7, one can see fluorescence spectra of matured and undeveloped cells of leaves from the medicinal species *Urtica dioica* and female "cones" of *Humulus lupulus*. The maxima 467–475 nm (flavonoids) increased, and the 540 nm shoulder (perhaps carotenoids and flavins) disappeared in adult plants of *U. dioica*. The developing glands of *H. lupulus* had a peak 430–450 nm, specific to

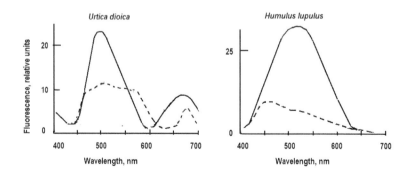

FIGURE 2.7   Estimation of plant material (leaves of *Urtica dioica* and scales of female cones of *Humulus lupulus*) readiness for phytopreparation. Mature cells—solid lines, imma-ture cells—dashed lines.

sesquiterpene lactones (Roshchina 2008), and a shoulder at 530 nm (flavins), while matured cones showed only one large maximum at 530 nm.

2.7.2 EXTERNAL FACTORS

Among external factors influencing autofluorescence are UV radiation, ionizing radiation, high doses of tropospheric ozone, microbial invasion, and temperature (Roshchina V.V. and Roshchina V.D. 2003; Roshchina 2008, 2014). Light may be an important factor for medical plant cultivation. For example, the influence of sunshine hours on azulene synthesis in *Matricaria chamomilla* L. has been shown to be greater than that of soil alkalinity (Singh et al. 2011). The effects of ionizing and gamma irradiations on plant cell autofluorescence are known only from the few experiments on pollen (Roshchina et al. 1998b). Light quality and intensity, moistening, pH, and metals, as well as ozone and its derivatives, both free radicals and peroxides/ozonides, influence autofluorescence (Figure 2.8). This is described in detail in earlier publications: Roshchina et al. 1998a; Roshchina and Roshchina 2003; Roshchina 2008). Here, attention has been paid mainly to examples of medicinal plants.

2.7.2.1 Ozone and Ageing

Among the secretory products from medicinal plants, the most significant changes occurred after ozone treatment, which often induces earlier aging of plants (Roshchina and Roshchina 2003). Tropospheric O_3 is formed in air as a result of industrial and automobile pollution undergoing UV-irradiation (Roshchina and Roshchina 2003: Cathokleous et al. 2015). UV light is not only the origin of ozone in nature but also contributes to cellular damage and aging. The cuticular wax of plant cells may absorb UV-A (<290 nm) and UV-B (290–320 nm) light, protecting plants from hazardous UV-irradiation (Jacobs et al. 2007). Ozone acts on bioaerosols containing pollens, fungi, bacteria, and biothreat agents, changing their fluorescence spectra (Santarpia et al. 2012). The aerosols appear to precipitate on plants collected for the pharmaceutical industry, and the admixture fluoresces on UV excitation.

The storage time of raw plant materials may be of important significance for pharmacy, because lipofuscin, the pigment of ageing, is formed in both vegetative organs (Brooks and Scallany 1978; Merzlyak 1988, 1989; Merzlyak and Zhirov 1990) and generative cells such as pollen of various species (Roshchina and Karnaukhov 1999). The products of lipid oxidation and peroxidation may decrease the quality of phytopreparations. Figure 2.9 shows the effects of ozone on the autofluorescence of plant materials after treatment of whole plants *Plantago major* and *Saintpaulia ionantha* with ozone doses of 0.05 and 0.005 μl/L, respectively. In this case, the formation of the fluorescent products at lipid oxidation/peroxidation or transformation of phenolic compounds takes place. Especially sensitive to ozone were leaves of the medicinal plant *P. major* (widely used in the care of gastrointestinal diseases), where yellow fluorescence was observed after the ozone treatment that was absent in untreated samples. Anthocyanin-containing cells from violet-blue petals of flowers from the African medicinal plant *S. ionantha* Wendl were considered as models for the analysis of metabolic damage under the action of various concentrations of ozone (Roshchina et al. 2017a). Laser-scanning confocal microscopy enabled observation

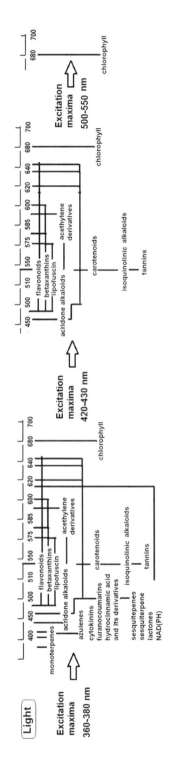

FIGURE 2.8 External factors influencing plant autofluorescence.

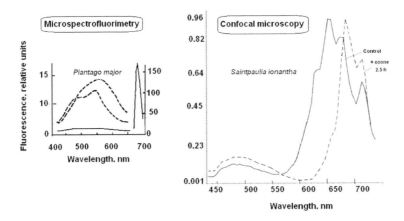

FIGURE 2.9 Effects of ozone on autofluorescence of plant materials after treatment of whole *Plantago major* and *Saintpaulia ionantha* plants with an ozone dose of 0.05 and 0.005 µl/L, respectively. Measurement of autofluorescence of leaves *P. major* and petals of *S. ionantha* was done by microfluorimetry and confocal microscopy. Control—solid line, ozone—dashed lines. (From Roshchina et al. 2017a and unpublished data.)

of the changes based on different images of autofluorescence and the fluorescence spectra of individual cells as well as of the surface tissue cells and surface secretory hairs. Low doses of ozone 0.005 µl/L (2.5 hours' exposure) or 0.008 µl/L (4 hours' exposure) do not affect or only weakly stimulate the autofluorescence of the samples. Acute experiments during 25–50 h treatment with O_3 at doses of 0.05–0.1 µl/L led to alterations in the position of maxima; in particular, peaks at 630–650 nm, specific to anthocyanins, disappeared from the fluorescence spectra of the main cells of petals. The chlorophyll maximum 675–680 m was seen for up to 25 h of exposure and disappeared by 50 h of ozonation. In the secretory hairs, a similar picture was seen in some cells of the stalk, whereas the head of the trichome, distinct from the other main cells, fluoresced with a maximum of 480–500 nm. Flavonoids and polyphenols were visible fluorescent components (Kopach et al. 1980), which formed or were intensively accumulated at the stress. The first alterations were fixed in the emission spectra of some petal cells after just 2.5 h exposure to O_3 (total dose 0.005 µl/L). This was a disappearance of maxima 620–630, 640–650 and 665 nm peculiar to the characteristic of anthocyanins, although no changes in the fluorescence images had been seen yet. Acute experiments during 25–50 h exposure to O_3 (doses 0.05–0.1 µl/L) led to changes in common fluorescence images (quenching of the emission in main cells of the petal surface and stalks of the secretory hairs). This quenching was especially clear at 620–660 nm, related to anthocyanins, whose emission in this range disappeared in the main petal cells. In the secretory hairs, a similar picture was seen in some cells of the stalk, whereas the head of the trichome, distinct from main cells, fluoresced with maximum in range 480–500 nm. Here, additional emission in the yellow-orange region was also seen. Experiments with individual phenols and chlorophyll treated with ozone showed that only anthocyanins in vacuoles and chlorophyll in chloroplasts are targets of O_3 even at the lowest concentrations

(Roshchina et al. 2017a). The sensitive cells of flowers are recommended as possible ozone bioindicators both outdoors and indoors.

In Table 2.3, the effect of ozone on the maxima position of dry individual compounds contained in *Saintpaulia* petals is shown. Among the individual substances treated with ozone at room temperature (22 °C), anthocyanins are most sensitive (Roshchina et al. 2017a). Fluorescence from the sum of *Saintpaulia* anthocyanins is lost even earlier than is seen for pelargonidin alone. The emission, peculiar to red anthocyanin pelargonidin (at pH < 7), completely disappeared after the use of the threshold concentration 0.1 µl/L ozone. A similar picture is observed for its analog, the blue color cyanidin (at pH > 7). Cyanidins, the form of anthocyanins at pH > 7, in water solutions have maxima at 610–630 nm at 220–230 or 270–280 nm excitation (Rakić et al. 2014). In literature, the fluorescence of compounds such as malvidin and its glycosides have maxima in blue-green up to 520 nm and in red 610–630 nm in acid/water solutions (Figueiredo et al. 1990; Lima et al. 1994; Drabent et al. 1999). Anthocyanins may fluoresce not only in blue; some of them emitted with maxima in the orange-red, 580–620 nm, too (Singh and Mishra 2015). According to investigations with secretory cells (Roshchina 2008), in longer spectral regions, besides green fluorescence (at 510 and 515 nm), yellow, orange, and even red (peaks 555, 570–585, and 610 nm) emission has been seen. For example, under UV microscope light, yellow fluorescence of the anthocyanin pelargonin (aglycon) and its glycoside

TABLE 2.3
Effects of Ozone on the Fluorescence Maxima of Individual Compounds Dissolved in 96% Ethanol (Excitation 360 nm)

	Maximum, nm/Fluorescence intensity changes			
Individual		**+ Dose of ozone (µl/L)**		
compound	**Control**	**0.05 (25 h)**	**0.1 (50 h)**	**0.2 (100 h)**
Sum of anthocyanins from *Saintpaulia* petals (pH 7.8)	430, 630–650	430/decrease	0	0
Anthocyanin pelargonidin (pH 5.5)	450–460, 665	450–460	410–420/decrease	0
Chlorophyll	675–680	675–680[a]	675–680/increase	675–680/decrease
Caffeic acid	450–460	450–460[a]	450–460[a]	450–460[a]
Ferulic acid	450–460	450–460[a]	450–460[a]	450–460[a]
Cinnamic acid	450–460	450–460[a]	450–460[a]	450–460[a]
Esculetin (Aesculetin)	500–520	450[a]	450/decrease	450/decrease
Coumaric acid	450	450[a]	450[a]	450[a]

[a] No changes in emission intensity.

pelargonidin has been observed. Pelargonidin in 96% ethanol (pH < 7) has emission maxima in blue. Anthocyanin fluorescence depends on the formation of complexes or conjugates with metals or other cellular components. When the compounds are applied to silica gel plates or Avicel ST microcrystalline cellulose, delphinidin, petunidin, and malvidin fluoresce in blue, green, and red, while cyanidin and peonidin fluoresces in red and pelargonidin in the orange-red spectral region (Wolfbeis 1985). In rigid, crystalline-state flavonoids, although most of them also fluoresce in blue at 450–470 nm, flavones also may have a maximum of 580 nm, but after fumigation with ozone (dose 5 µl/L), a shift was observed from the blue to the orange spectral region (Roshchina and Melnikova 2001; Roshchina 2008).

Chlorophyll is not too sensitive in comparison with anthocyanins—its maximum at 675–680 nm decreased only after 50–100 h (dose 0.1–0.2 µ/L) ozone action (Table 2.3). Unlike chlorophylls and anthocyanins, chlorogenic, ferulic, and caffeic acids, as well as coumaric acid dissolved in ethanol, may fluoresce in blue with maxima at 450–460 nm, and the emission intensity did not change during acute exposure (Table 2.3). Thus, strong yellow emission in the head of the secretory hair on the *Saintpaulia* petals after 50 h ozonation may be due to the above-mentioned phenolic acids and coumaric acid. The emission of the coumarin esculetin at 450 nm under higher doses of O_3 decreased. 7-Alkoxy coumarins generally have a purple fluorescence, whereas 7-hydroxy and 5,7-dihydroxycoumarins tend to fluoresce in blue (Wolfbeis 1985). In the dry state, o-coumaric acid emits in blue with maxima at 440–450 nm and esculetin at 460 nm (Roshchina and Melnikova 2001; Roshchina 2008). This does not exclude the possibility that new phenols may also be formed under acute ozone doses.

The effects of ozone on living systems, both animals and plants (Roshchina and Roshchina 2003), and the chemical mechanisms of the action (Calvert et al. 2015) is at the center of attention when dealing with the biggest danger for humanity: ground-level O_3. Organisms are sensitive to high concentrations of ozone (higher than 0.1 µl/L per hour, at which there is a characteristic odor) or chronic levels of this gas (lower than 0.05 µl/L) in later cases one can indicate early reactions, characteristic for damage at the cellular level. Different concentrations of ozone cause various reactions, from simulating processes at low doses (< 0.1 µl/L) to a breaking of the growth and appearance of necrotic and age spots at high doses (in the range of 0.1–2 µl/L).An intensive search is ongoing for ozone indicators such as simple model systems suitable for environmental biomonitoring and showing the stress induced by ozone (Manning and Feder 1980; Roshchina and Roshchina 2003).

Pollen ability to germinate at the exposure of various concentrations of ozone has been shown in some examples: maize *Zea mays* (Mumford et al. 1972), plantain *Plantago major* (Roshchina and Melnikova 2001; Roshchina and Roshchina 2003; Roshchina et al. 2003; Pasqualini et al. 2011) as well as vegetative microspores of horsetail *Equisetum arvense* (Roshchina and Melnikova 2000). Moreover, it has been shown that high concentrations of ozone increased the allergenicity by birch pollen (Beck et al. 2013). The accumulation of stress metabolites, such as biogenic amines – neuromediators (Roshchina 1991; Roshchina 2001) also occurs under the influence of O_3 (Roshchina and Roshchina 2003).

Among the targets of ozone in the studied plants were pigments such as chlorophyll, anthocyanins, carotenoids, and azulenes, summarized in monographs

(Roshchina and Roshchina 2003). Stress products, such as catecholamines and histamine, or pigments of aging (lipofuscin) were also indicators of damage. Mainly, experiments have been carried out using chronic exposures: ozone treatment + rest, then again ozone treatment + rest. Less attention was paid to acute exposures, when the stress is most intensive.

The leaves of *Urtica dioica* are valuable due to their high concentration of chlorophyll, used in the phytopreparation "Chlorolipt" in Russia. However, one must need to keep in mind that the emission of chlorophyll maximum 680–685 nm excited by light 430 nm is changed first in acute experiments with ozone (non-discrete exposure). It is an indicator of the pigment damage as shown by microspectrofluorimetry in Table 2.4. After 2.5 h ozonation, in *U. dioica* leaves, yellow-green autofluorescence appeared, whereas red emission (related to chlorophyll) decreased both in the hairs and in the nonsecretory part. Yellow crystals (absent in the control) appeared, like those shown for radish leaves (Roshchina and Roshchina 2003; Roshchina et al. 2016b).

When the concentration of tropospheric ozone is high (along autostradas, freeways and autoparkings in parking lots, or industrial pollution, etc.), the collection of herbs for pharmacy may lead to negative effects for humans due to the risk of many free radical processes taking place in the presence of the oxidant. Recently, changes in the autofluorescence of pollen have also been considered as a sensitive indication of ozone activity. It has been demonstrated that the pigments contained in the surface of a pollen grain determine its ability to resist O_3 (Roshchina and Karnaukhov 1999; Roshchina and Melnikova 2001). Pollen collected from natural plants, for instance pines and meadow grasses, as well as from artificial decorative species may change the color of its fluorescence under conditions of chronic ozone fumigation. This change is especially marked when the fluorescence maximum shifts from the yellow and orange regions of the spectrum to shorter-wavelength regions, an example of which has been shown in a monograph (Roshchina and Roshchina 2003).

Shifts of the maxima in the fluorescence spectra of pollen, as registered by the method of microspectrofluorimetry, indicate that a degree of damage has been caused by ozone. Pollen lacking the appropriate pigments in its surfaces is injured more severely (Roshchina and Roshchina 2003).

TABLE 2.4
Effect of Ozone on the Fluorescence Intensity I_{680} (in Relative Units) of Leaf Cells in *Urtica dioica* in Acute Experiments

Excitation	Control	Dose of ozone		
		2.5 h (0.005 μl/L)	50 h (0.l μl/L)	75 h (0.l5 μl/L)
Excitation by light 360–380 nm	1.79 ± 0.01	2.11 ± 0.04	0.14 ± 0.009	0.18 ± 0.010
Excitation by light 430 nm	7.85 ± 0.20	1.31 ± 0.10	0.6 ± 0.41	0.17 ± 0.5

The effects of temperature on grown medicinal plants are rarely studied. In leaves of *Saintpaulia* sp., the cold induced collapse in vacuoles in palisade cells (Kadohama et al. 2013). It is well known that Saintpaulia leaf is damaged by the rapid temperature decrease when cold water is irrigated onto the leaf surface. This temperature sensitivity and the mechanisms of leaf damage have been studied in Saintpaulia (*Saintpaulia* sp. cv. "Iceberg") and other Gesneriaceae plants (Kadohama et al. 2013). Saintpaulia leaves are damaged and discolored when subjected to a rapid decrease in temperature, but not when the temperature has decreased gradually. Sensitivity to rapid temperature decrease was enhanced within 10 to 20 min during pre-incubation at a higher temperature. Injury was restricted to the palisade mesophyll cells, where there was an obvious change in the color of the chloroplasts. During a rapid temperature decrease, chlorophyll fluorescence, monitored by a pulse amplitude modulated fluorometer, diminished and did not recover even after rewarming to the initial temperature. Isolated chloroplasts were not directly affected by the rapid temperature decrease. Intracellular pH was monitored with a pH-dependent fluorescent dye. In palisade mesophyll cells damaged by rapid temperature decrease, the cytosolic pH decreased and the vacuolar membrane collapsed soon after a temperature decrease. In isolated chloroplasts, chlorophyll fluorescence declined when the pH of the medium was reduced. These results suggest that a rapid temperature decrease directly or indirectly affects the vacuolar membrane, resulting in a pH change in the cytosol that subsequently affects the chloroplasts in palisade mesophyll cells that further confirmed that the same physiological damage occurs in other Gesneriaceae plants.

Besides the changes in the chlorophyll affected by ozone, this fluorescence method permits researchers to measure the shifts in the maxima emitted by cell surfaces in the light-blue and orange regions of the spectrum. The tropical species *Astronium graveolens* (family Anacardiaceae), a medicinal plant used in officinal or folk care in Mexico and Central and South America as antioxidant antibacterial plant (Hernandez et al. 2012; Bandeira 2018), is found to be sensitive to ozone. Studies by Brazilian scientists (Fernandes et al. 2016) have shown brown stipplings in the leaf intercostals in response to chronic and acute ozone exposure. Lysigenous secretory ducts and palisade parenchyma in similar samples fluoresced in the 500–540 nm (polyphenols) and > 650 nm (chlorophyll) regions due to the changes in oxidative metabolism. Flavonols, anthocyanins, and condensed tannins stand out as polyphenols usually involved in ozone detoxification.

The damage induced by O_3 can be linked to a change in the composition of phenolic and terpenoid compounds, as both have double bonds susceptible to oxidation, along with the formation of lipofuscin (Roshchina and Roshchina 2003; Roshchina 2008). Thus, the analysis of the light emissions from plant surfaces, which are therefore unrelated to chlorophyll, can also serve as an early indicator of ozone damage.

The autofluorescence of secretory structures such as glandular cells and hairs is also changed in response to ozone treatment (Roshchina and Roshchina 2003). In particular, this modification is readily visible in secretory cells containing lipid-like secretions that fluoresce in the blue-green or yellow region. Among the components are pigments of aging—lipofuscins—as demonstrated on the pollen of various plants and vegetative microspores of horsetail (Roshchina and Karnaukhov 1999; Roshchina V.V. and Roshchina V.D 2003; Roshchina et al. 2015).

A defensive barrier in the response to ozone is the cell wall, whose antioxidant characteristics deal with the presence of phenols in the polymer matrix. For example, exine of pollen grains is able to inactivate free radicals and hydrogen peroxide (Smirnova et al. 2012). Autofluorescence of the isolated pollen exine of *Nicotiana tabacum* (sort Petit Havana SR1) increased from 440 to 460 nm, depending on pH, and the presence of ferulic and sinapic acid in the polymeric matrix as well as large amounts of hydroxycinnamic acid esters has been shown. These phenolic compounds appear to play a key role in scavenging free radicals and peroxides.

Depending on the urban situation, the collection of plant material for the pharmaceutical industry may encounter difficulties because of damage to leaves and flowers. Very marked changes visible in cities and towns as well as in suburbs have been demonstrated in a publication by Bharti with co-authors (2017). Research in this direction is necessary for the industry.

2.7.2.2 Parasitic Invasion

Another damaging factor influencing autofluorescence is foreign invasion. Insect or microbial attacks may be visible simply under a luminescence microscope (Table 2.5). For pharmacy, it is necessary to know the main indicators of parasitic invasion if using the fluorescent method (Rohringer et al. 1977). Luminescence microscopy of plants has shown weak autofluorescence of aphids (Roshchina 2012). According to

TABLE 2.5
Examples of Autofluorescence of Parasitic Invasion Cells in Plant Materials

Fluorescence in plants	Maxima, nm			
	Blue	Green	Yellow-Orange	Red
Aphids	475	540		675–680 (eat leaves with chlorophyll)
Western corn rootworm (WCR; *Diabrotica virgifera virgifera*) is the most significant pest	473.6 (in cortical tissue)	557.9 yellow/ green (dominant)		
Cereal cyst nematode	487.6			
Arbuscular mycorrhizal fungi: fluorescent arbuscules in the cortex			583.9 nm (yellow)	
Plant and soil fungi *Fusarium-*				602 nm
Mycorrhizal fungi *Glomus*: hyphae, vesicles, and arbuscular branches of species			Yellow	

Sources: Jabaji-Hare, S.H., et al., *Canadian Journal of Botany-Revue Canadienne De Botanique*, 62, 2665–2669, 1984; Wolfbeis, O.S., *Molecular Luminescence Spectroscopy. Methods and Applications*, Wiley, New York, Chichester, Brisbane, 1985; Roshchina, V.V., *Int. J. Spectrosc.*, ID 124672, 1–14, 2012; Donaldson, L. and Williams, N., *Plants (Basel)*, 7, E10, 2018; Galindo-Castañeda et al., *J. Exp. Bot.*, 70(19), 334–344, 2019; Strock et al., *J. Exp. Bot.*, 70(19): 5327–5342, 2019.

many authors (Karnaukhov 1978; Monici 2005; Roshchina 2012; Donaldson and Williams 2018), at excitation >350 nm animal cells demonstrate fluorescence in the visible region due to collagen (max. 400 nm), elastin (max. 420 nm), NADH (max. 460 nm), flavins (max. 520–530 nm), lipopigments (540–550), and porphyrins (max. 630–640 nm), while cutins are not known to be autofluorescent under UV light. Often, porphyrins have been observed in necrotic tumors and are well differentiated from chlorophyll in plant tissues (Monici 2005).

Western corn rootworm (WCR; *Diabrotica virgifera virgifera*), the most significant pest affecting maize across North America and Europe, causes production losses, and pest management costs in the USA alone exceed US$1 billion per annum (Strock et al. 2019). The first instar larva was ~2 mm in length, had a diameter of 0.42 mm, and was observable entirely within the cortical tissue of the maize root. The fluorescence from the WCR larva was yellow/green in color and had a dominant emission wavelength of 557.9 nm, while cortical tissue fluoresced blue with a dominant emission wavelength of 473.6 nm (Strock et al. 2019).

Recently, laser ablation tomography (LAT) of roots was studied across a diversity of plant species and edaphic organisms as a result of the contrasting autofluorescence spectra received from different tissues (Galindo-Castañeda et al. 2019; Strock et al. 2019). In Table 2.5, some data for fluorescence in the presence of parasites are shown. Most parasites, for example aphids, weakly fluoresce in blue—420–430 nm. Bright fluorescence in blue-green usually is peculiar to the cell surface of fungi, nematodes, and some arthropods. Often, we can see fluorescence changes in the plant tissue itself induced by parasitic infection. Blue fluorescence may be due to cell wall cellulose (420–430 nm), suberin (465 nm), ferulate (480 nm), or chitin of insects (420–430 nm) as well as plant excretions and secretions—terpenes and some flavonoids (430–470 nm). Red fluorescence of *Fusarium*-damaged tissue may be caused by secondary metabolites, presumably the phenol products produced in response to fungal invasion. In observations of intact cortical tissue in maize, barley, and common bean roots (Galindo-Castañeda et al. 2019), blue autofluorescence from plant cell walls under UV excitation was seen due to lignin (455 nm) and additionally blue/green spectra (470–525 nm). The orange fluorescence of stele tissue maize (588.6 nm) may be a product of anthocyanins, azulenes, polyacetylenes, isoquinoline and acridone alkaloids, which are also known to emit in yellow/orange/red. If the roots of medicinal plants contain most of the active compounds in their autofluorescence, perhaps there is a contribution due to microbial relations. Root cell walls are primarily composed of lignin, cellulose, hemicellulose, suberin, and tannins (Cohen 1987; Foster et al. 2005; White et al. 2011). Additionally, the epidermis of nematodes is composed of a thick cuticle made of collagen (Anya 1966). Overlapping with our spectral observations of cereal cyst nematode (487.6 nm) and WCR (557.9 nm) from LAT (Table 2.5), under UV excitation, aphids display maxima of 475, 540, and 675–680 nm (Roshchina, 2012). As shown by laser ablation tomography (LAT), arbuscules, formed by arbuscular mycorrhizal fungi in the cortex, fluoresced in yellow (583.9 nm), similarly to observations by Jabaji-Hare *et al.* (1984) where hyphae, vesicles, and arbuscular branches of *Glomus* species emit bright yellow under UV excitation (Table 2.5). Red fluorescence of *Fusarium*-damaged tissue (602 nm) may be the product of secondary metabolites (Table 2.5) produced in response to fungal invasion, such as phenols, which have been observed to emit under UV excitation (Wolfbeis 1985). Moreover, plant tissue damaged by reactive oxygen species may form lipofuscin pigments, which mainly fluoresce in blue (Merzlyak 1988, 1989; Merzlyak and Pogosyan 1988; Merzlyak and Zhirov 1990), although some of them emit in the red range >600 nm (Roshchina 2012). Gange and coworkers (1999) indicate that arbuscular mycorrhizal fungi are unique among a broad range of fungal taxa in their capacity to fluoresce under UV light. Unlike arbuscular mycorrhizal fungi, fluorescent structures of *Fusarium* were not visible with the laser ablation tomography. The absence of autofluorescence under UV excitation in some pest organisms has been previously reported (Ames et al. 1982). Variance in the dominant spectra may reflect differences in the host and root symbiont species.

Overground plant tissues also may demonstrate the presence of parasites on the surface and show different fluorescence of the host and the invader. On Figure 2.10, one can see the appearance of butterfly grena with eggs on the leaves of the medicinal species *Urtica dioica*. Instead of red fluorescence of the cells, the insect invasion forms a plateau (secretions fluoresce in blue, and red fluorescence is weak or even not seen) where the eggs stick and start to develop up to their metamorphosis into caterpillars. The orange (in transmitted light) interior of grena with eggs fluoresces weakly within the brightly fluorescent environment of the damaged leaves.

Earlier, confocal microscopy was applied to study fungal infection, although using dyes (Kwon et al. 1993.). Autofluorescence related to fungal infection has been shown using a luminescence microscope, for example with medicinal plants such as *Callisia fragrans* (Lindl.) Woodson (family Commelinaceae), whose leaves are used to treat rheumatic diseases (Roshchina 2014). Brown necrotic spots appeared much later, after a researcher had been able to see blue fluorescence peculiar to fungal invasion. In another case, the formation of color flavonoids as signal molecules served as visual diagnostics (Peer and Murphy 2006).

2.7.2.3   Water and Salt Stresses

Autofluorescence of cellular compounds may be used as a marker for cytodiagnostics of medicinal plants undergoing water and salt stress during their development. It may also be considered as a natural indicator of the cellular state, because its appearance indicates altered cellular metabolism and is a response to abiotic stress (Jiménez-Becker et al. 2019. This phenomenon has been clearly observed in plant secretory structures analyzed using a luminescence microscope.

In fluorescent microscopic studies, highly autofluorescent compounds were observed in both cotton seed genotypes of cotton under water stress (Kolahi et al. 2019). The structures are not visible utilizing a conventional microscope without special histochemical staining. But slightly dried samples fluoresce in blue, which shows the influence of water stress.

Salinity is a major abiotic stress limiting the growth and productivity of plants in many areas of the world due to the increasing use of poor-quality water for irrigation and soil salinization (Gupta and Huang 2014). The typical definition of a halophyte is a plant species that can survive and reproduce under growth conditions with more than 200 mM NaCl. It has been conservatively estimated that there is approximately 1000 million ha of salt-affected land throughout the world (Liu et al. 2018). Salt stress is a combination of ionic stress due to the chaotropic effects of incoming Na^+ and Cl^- and osmotic stress resulting from a decrease in water potential (Lugan et al. 2010).

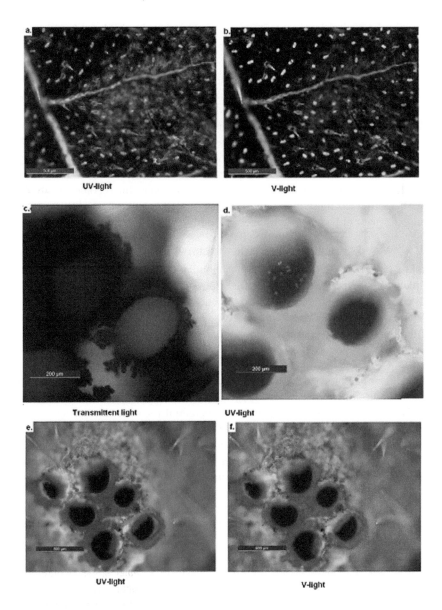

FIGURE 2.10 Autofluorescence of the damaged surface of *Urtica dioica* leaves. (a) and (b) earlier stage of the invasion of butterfly *Vanessa* sp. seen under luminescence microscope in UV light (360 nm) or in violet light (430 nm); (c)–(f) visible developing grena with eggs.

The result of these disturbances in water management is a loss of turgor, inhibition of cell elongation, stomatal closure, and decrease in the intensity of photosynthesis (Cassaniti et al. 2012). Therefore, it is important to carry out studies that allow the selection of suitable, salt-tolerant species. Some medicinal plants have the ability to grow under salinity due to the presence of different mechanisms for salt tolerance; such plants are known as salt-resisting plants, salt-tolerating plants, or halophytes (Aslam et al. 2011).

Salt glands, as the halophyte adaptation of similar plants to grow, fluoresce at 440–470, 530–540 or 565 nm (Roshchina 2008). The blue and yellow fluorescence may reflect the presence of phenols (for example, flavonoids), which can be histo-chemically confirmed by the iron chloride reaction.

As naturally salt-tolerant plants, halophytes can grow in a variety of saline habitats due to the development of special adaptations, particularly secondary metabolites with antioxidant properties (Stevanovic et al. 2019). Since, in order to overcome harsh environmental conditions, halophytes have the ability to produce phenolic molecules with powerful biological capacities, this interesting ecological group of plants has attracted more attention in recent years because of a rapid increase in demand for natural bioactive substances.

2.7.2.4 Metal-Enriched Soil as Stress Factor

Metal by-products of industrial activity accumulating in soil are able to change normal plant metabolism. This is the cause of possible disturbances in the pharmaceuticals contained in medicinal plants grown on such soil (cultivated or collected), and autofluorescence may be a tool for diagnostics. One example is aluminum resistance leading to the continuous excretion of organic acids (Zheng et al. 1998).

2.7.2.5 Fluorescence in Monitoring of Bioparticles and Agrosystems

The monitoring of bioparticles and plants in agrosystems based on fluorescent methods is an environmental task (Pan et al. 2007, 2011). As early as 1999, Cerovic and coworkers presented a paper related to the potential for such monitoring for cultivated plants, measuring the red fluorescence of chlorophyll and the blue-green emission of ferulic acid and its derivatives excited by UV light.

One type of pollution is the presence of biological particles in the air, which may be observed later on the surface of medicinal plants as visible pollution or may be detected as an aerosol. Primary biological aerosol particles can significantly influence climate and precipitation systems as cloud nuclei and can spread disease to humans, animals, and plants (Healy et al. 2014; Pan 2014). These particles originate from plants (pollen grains of seed-bearing species or spores of spore-bearing species), microbial spores of microorganisms, and their fragments. Using air probes, researchers can detect these particles if they fluoresce (Huffman et al. 2010; Pöhlker et al. 2011; Pan 2014; O'Connor 2013, 2015; Healy et al. 2014; Fennelly, M. J. et al. 2018). A waveband integrated bioaerosol sensor (WIBS-4) with a UV aerodynamic particle sizer (UV-APS) were compared with real-time fluorescence techniques resulting in the fluorescence and optical microscopy of impacted samples (Healy et al. 2014). Both real-time instruments showed qualitatively similar behavior, with increased fluorescent bioparticle concentrations at night, when relative humidity was highest and temperature was lowest. According to O'Connor and coworkers (2013, 2015), who used confocal microscopy and real-time instruments such as time-resolved confocal microscopy and Wave-band Intergrated Bioaerosol Sensor (WIBS), individual pollen grains of *Dactylis glomerata*, *Lolium perenne*, *Anthoxanthum odoratum*, *Fagus sylvatica*, and *Quercus robur* emitted in the spectral range 450–650 nm, and if only developed but not matured, had a maximum of 675–680 nm in the fluorescence spectra, which is specific to chlorophyll. Ascospores of *Cladosporium* were also studied. The apparatus permits both qualitative and quantitative measurements. Real-time fluorescence instrumentation to monitor biological particles (ascospores, basidiospores, and *Ganoderma* spp.) has been used by Fennelly et al. (2018).

2.7.2.6 Xenobiotics

Among xenobiotics in plant cells, fluorescent pesticides may be found; for example, pesticidal carbamates acting on cholinesterases of both animals and plants (Addison et al. 1977). In addition, carbamates offer the possibility of determining their influence on cholinesterase activity. As early as 1992, Mueller and coworkers presented a list of fluorescent herbicides that could potentially be used in diagnostics. The fluorescence properties of 39 herbicides representing several major types of chemistry were determined. The fluorescence of analytical standards was measured in acetonitrile, acetonitrile + water, and acetonitrile + water + strong acid. Fourteen of the 39 herbicides fluoresced to some extent, and seven (bentazon, chloramben, difenzoquat, fluometuron, imazaquin, 2-methyl-4-chlorophenoxyacetic acid [MCPA], and norflurazon) were identified as good candidates for further method development. Herbicides or their derivatives have been detected with spectrofluorometric methods in various matrices, including fluometuron and its metabolites in soil to 20 ng/g soil, asulam (methyl[(4-aminophenyl)sulfonyl]carbamate) in spinach, and glyphosate ((N-phosphonomethyl) glycine) and its metabolite in natural water. This technique has also been employed to identify other pesticides, such as the methylcarbamates in food, carbaryl (1-naphthyl-*N*-methylcarbamate) in honeybees or honey, and thiabendazole and carbendazim in various crops. The bipyridiniums fluoresced in acetonitrile very strongly. However, addition of water to the solutions of diquat and paraquat totally quenched the observed fluorescence. Additionally, diquat and paraquat are only very slightly soluble in acetonitrile (but very water soluble), and this would hinder method development. It should be noted that this method has not so far been used for histochemical analysis in plant cells *in situ*. Only chlorophyll fluorescence is recommended for analysis as a marker for herbicides (Dayan and de Zaccaro 2012).

Recently, the pesticide glyphosate was localized after treatment with it (Cardoso-Gustavson et al., 2018). But this pesticide has no fluorescence. Since glyphosate does not exhibit fluorescence per se, an indirect approach was considered for its route inside leaf tissues, such as an inert tracer or some specific alteration caused by this herbicide. An interesting approach would be the use of something that could indirectly follow glyphosate by direct tracking. Since glyphosate increases cell permeability, Lucifer Yellow (LYCH; 6-amino-2,3-dihydro-1,3-dioxo-2-hydrazinocarbonylamino-1H-benz[d,e]isoquinoline-5,8-disulfonic-acid dilithium salt) is recommended as an apoplastic tracer of glyphosate uptake. The dye has a very high quantum yield, resulting in easily visualized fluorescence after uptake by cells; the small size of the LYCH fluorescent molecule allows passage through the cell wall; it is stable over a wide range of pH, from 2.0 to 10.0; and it is not toxic to cells. Such microscopic markers might shed light on the recognition of glyphosate-resistant and susceptible plant species or varieties, which may vary in C3 and C4 species due to their putative distinct leaf anatomy, reaching the vascular bundles of sensitive species and becoming retained in the leaf tissues of resistant ones. All species exhibited a decrease in chlorophyll content at the site of glyphosate application regardless of their photosynthetic metabolism or susceptibility. Glyphosate-sensitive species showed alterations in chloroplast morphology and activity of non-enzymatic antioxidants (carotenoids and flavonoids). In addition to symptoms indicating a process of accelerated cell senescence, a specific type of cell necrosis (hypersensitive response) was observed in resistant C4 species as a way to prevent the translocation of this herbicide, while resistant C3 species accumulated phenolic compounds inside the vacuole, probably sequestrating and inactivating the glyphosate. Specific autofluorescence from 370 to 700 nm, exciting at 364 nm, appeared from different cells or intracellular structures: flavonoids usually exhibit a strong emission between 400 and 500 nm, carotenoids, between 500 and 550 nm, and chlorophyll, > 650 nm (Hutzler et al. 1998; Roshchina 2008; Fernandes et al. 2016). The results imaging in a degradation may be due to a production of fluorescent by-products (lipofuscins and pheophytins, respectively [Roshchina, 2008]) that can be identified by confocal microscopy but neither of them was observed here. The other probable reason, rather than supporting the hypothesis of suppression of chlorophyll accumulation, could be that our observations were taken prior to pheophytin formation. This is plausible, since chlorophyll degradation is highly regulated. The causes and implications of the reduction of chlorophyll emission are related to the cell signaling mediated by reactive oxygen species (Maroli et al. 2015). After glyphosate application, the herbicide induces oxidative bursts initially in the mitochondria (Gomes and Juneau 2016) and then in chloroplasts, where it blocks the shikimate pathway (Gomes et al. 2017). Both events generate reactive oxygen species that may act as signals and/or cause cell damage.

How herbicides influence the content of flavonoids in various weed species of Poaceae has been shown by leaf autofluorescence (Hjorth et al. 2006). Glyphosate or iodosulfuron reduced the levels of chlorophyll and the flavonoids quercetin, naringenin, and/or naringin. Prosulfocarb acted mainly on the flavonoid concentration, increasing the height of the 527 nm maximum in the autofluorescence spectrum.

2.8 NATURAL STRESS COMPOUNDS FORMED IN PLANTS

One result of unfavorable growth conditions for medicinal plants may be an increase in fluorescent components that are lacking or completely absent under normal conditions. Usually, salt, ozone, feeding by insects or other causes may induce the formation of these components, which serve as a marker of damage in pharmaceutically valuable material.

Flavonoids and polyphenols are visible fluorescent components that are formed or intensively accumulated under stress and influence the quality of phytomaterials for pharmacy (Kumar and Pandey 2013). In many cases, this is a defense by plants. But also, the metabolites may be included in insect nutrition. Phenolic compounds were oxidized, hydrolyzed, or modified in various ways through the digestive tract of larvae (Vihakas 2014). High concentrations of complex flavonoid oligoglycosides were found in the hemolymph (the circulatory fluid of insects) of birch and pine sawflies. The larvae produced these compounds from simple flavonoid precursors present in the birch leaves and pine needles. Flavonoid glycosides were also found in the cocoon walls of sawflies, which suggested that flavonoids were used in the construction of cocoons. Salt in soil or ozone induces the formation of new phenol compounds, mainly flavonoids.

In a few experiments, the formation of fluorescent phenols has been observed in response to stress. The secretory hairs of *Saintpaulia ionantha* Wendl and anthocyanin-containing cells from the blue petals of flowers have been considered as models for fluorescent analysis (by luminescence microscopy, including the confocal microscopy technique) in the indication of damage arising under the action of various ozone concentrations (Roshchina et al. 2017a). Laser-scanning microscopy enabled the changes in images of autofluorescence and the fluorescence spectra of individual cells of petals and multicellular secretory hairs to be observed. The first alterations were in the emission spectra of some petal cells just after 2.5 h exposure to O_3 (total dose 0.005 µl/L)—the maxima at 620–630, 640–650, and 665 nm, specific to anthocyanins, were missing, although at an ozone concentration of 0.005 µl/L under a luminescence microscope, no changes in the fluorescence images were yet seen. Acute experiments during 25–50 h exposure to O_3 (doses 0.05–0.1 µl/L) led to changes both in common fluorescence images (quenching of the emission in the main cells of the petal surface and the stalks of the secretory hairs) and in the spectral position of maxima in the 620–660 nm region, specific to anthocyanins, that disappeared in the main petal cells. The chlorophyll maximum at 675–680 nm was seen for up to 25 h of exposure but disappeared by 50 h of ozonation. In the secretory hairs, a similar picture was seen in some cells of the stalk, whereas the head of the trichome, distinct from main cells, fluoresced with a maximum of 480–500 nm, and here, additional emission in the yellow-orange region was also seen. Experiments with individual phenols and chlorophyll treated with ozone showed that only anthocyanins in vacuoles and chlorophyll in chloroplasts are targets of O_3 at even the lowest concentrations. The sensitive cells of flowers have been recommended as possible ozone bioindicators both outdoors and indoors.

As well as flavonoids, salt in soil or ozone induces the formation of non-fluorescent substances in the visible spectral region (Korobova et al. 1988; Roshchina and Yashin 2014; Roshchina et al. 2015a, 2015b). Thus, large amounts of similar compounds are formed, such as histamine or catecholamines, that can be used as a preliminary estimate in histochemical procedures (Chapter 4). These biogenic amines are considered to be toxic in food and medicinal phytopreparations (Ekici and Omer 2019; Konovalov 2019).

It is also of interest to take a new look at the products of ozone and oxygen transformation, such as superoxide radical and peroxides, as well as natural redox agents occurring in plants: dopamine, noradrenaline, and adrenaline (Roshchina 1991, 2001). Superoxide is considered an obligatory, catalytic intermediate in the photosynthetic reduction of oxygen by adrenaline and dopamine (Allen 2003). The toxicity of the superoxide they produce has been suggested to rule out oxygen reduction as a physiological component of normal photosynthesis. In cell metabolism, adrenaline is transformed into adrenochrome in the presence of copper and iron. Adrenochrome has a red color. It can be converted into colorless fluorescing adrenolutin in a basic medium and in the presence of ascorbate. Adrenolutin 3,5,6-trihydroxy-1-methylindole is an intermediate in the metabolism of adrenaline and in the transformation of adrenochrome in melanins. The compound is known as one of the most strongly fluorescing metabolites. Its broad excitation spectrum and high quantum yields make this compound a potential effective acceptor of excitation energy (Polewski and Slawińska 1987). The characteristic yellow-green fluorescence develops rapidly when alkaline solutions of adrenaline stand in air. These catecholamines might function as chemical analogs of a proposed natural mediator, or oxygen-reducing factor (Allen 2003), that allows oxygen reduction to participate in energy transduction in photosynthesis. The fully oxidized form of adrenaline, adrenochrome, also acts as a mediator of photosynthetic oxygen uptake, but only by reducing oxygen to superoxide.

2.9 FLUORESCING PRODUCTS OF DAMAGE

Fluorescing products of cellular damage arise in response to many stress factors, such as ozone and reactive oxygen species or UV light, as well as due to aging. If the plant cell or tissue is damaged (aging, UV light, or ozone), the products of free radical processes, including lipid peroxidation, in living organisms can fluoresce (Halliwell and Gutteridge 1985; Gutteridge, 1995). The nature of fluorescing spots on the leaves and microspores of plants is believed to be related to Schiff bases formed by free radical reactions in the lipid fraction of membranes. Schiff bases arise when aldehyde groups of a substance react with amino groups of another compound. Principally, this may happen with any amino acid or amino acid residues in proteins. Aldehydes are formed in many metabolic processes, but under stress, malondialdehydes are often formed as a result of lipid peroxidation, which leads to the formation of fluorescent products with emission in blue (420–460 nm). Sometimes, they may emit in violet (400 nm), as is seen for Schiff bases formed by the reaction of glycolaldehyde with various animal proteins (Acharya and Manning 1983). Synthetic Schiff bases formed from different aldehydes (such as benzaldehyde and others) and amines (including derivatives of salicylic acid), when excited by light at 350–380 nm, emitted with maxima from 500 to 570 nm (Huisen and Xinfu 2003). The quantum yield of fluorescence (Q_f) of the Schiff bases, in particular that obtained from salicylaldehyde and 2-aminophen, is relatively high at nearly 0.78 (Santhi et al. 2004). The scheme of the appearance of the fluorescent products is presented in Figure 2.11. Deeper changes result in the occurrence of the fluorescing pigment lipofuscin, which

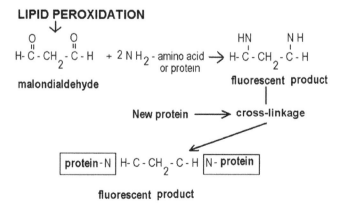

FIGURE 2.11 Common scheme of formation of fluorescent products due to stress and aging.

is frequently found in aging tissues of plants as a final product of lipid peroxidation (Merzlyak 1988). A contribution to the autofluorescence at 410–450 nm may also be made by lipofuscin. The fluorescence could be connected with the group of aging pigments, named as a whole lipofuscin, which are formed in animals from terpenoid compounds (Karnaukhov 1973). A chloroform-soluble product fluorescing with a maximum at 440 nm and similar to lipofuscin in animals was also found by Maguire and Haard (1975) in ripening fruits of banana (*Musa cavendishii*, var. Valery) and pear (*Pyrus communis*, var. Bartlett), mainly in peel and pulp tissue.

This pigment may be also derived from photosynthesizing cells (Merzlyak et al. 1984; Merzlyak 1988). In many cases, for example in corn samples, lipofuscin-like pigments fluorescing at 420–450 nm are formed under air pollution by ozone and nitrogen dioxide (Brooks and Csallany 1978). Evidence of the participation of lipofuscin in the orange-blue shift in the fluorescent spectra of intact pollen of *Philadelphus grandiflorus* under the ozone influence was provided by Roshchina and Karnaukhov (1999).

Ageing also induces lipid peroxidation and formation of fluorescent products (Roshchina 2003). For instance, the products of fresh and stored pollen differ, as seen from the fluorescence spectra of chloroform-ethanol extracts from pollen *Hippeastrum hybridum* and vegetative microspores of *Equisetum arvense*. Changes in their composition are seen from the fluorescence spectra of intact microspores (pollen and vegetative microspores) and their chloroform-ethanol (2:1) extract. More new fluorescent products are observed at ageing microspores compared with fresh microspores. Thus, analysis of autofluorescence could be an early indication of damage to the secreting cells induced by a stress factor.

2.10 CONCLUSION

The autofluorescence of plant secretory structures changes during their development and under the influence of various factors. It reflects the alterations in the composition or/and redox state of secretory products accumulated. The phenomenon could be an indicator of the appearance of secretions in secretory structures at earliest stages of development. Therefore, analytical techniques based on autofluorescence monitoring can be utilized in pharmaceutical practice.

Fluorescent analysis of plant secretory cells has the potential for cellular diagnostics both in fundamental studies and in all industrial conditions. Due to the possibility of obtaining rapid information in intact tissues and single cells without tissue homogenization procedures and long biochemical manipulations, luminescence observation of plant materials could be a non-invasive application to determine whether or not they are ready for pharmacy. This is possible because the main valuable drugs are concentrated in secretory structures. A second possibility deals with the estimation of cell viability, which is especially required for normal pollination and fertilization in genetic analysis for selection in agriculture and biotechnology. A third perspective use of fluorescence changes is analysis of cell damage, which is necessary for ecological monitoring of grown plant material for pharmacy. Earlier diagnosis of the disturbances induced by oxidative stress, aging, or pest invasion may be based on the fluorescence of secretory products as natural drugs.

3 Atlas of Autofluorescence in Plant Pharmaceutical Materials

It is necessary to show concrete applications of our knowledge of autofluorescence as a phenomenon to pharmacy based on practical use. Therefore, we have compiled a small atlas from our data gathered over the last 10 years.

3.1 *ACHILLEA MILLEFOLIUM*, COMMON YARROW

Common yarrow, *Achillea millefolium*, is usually used in medicine as a blood-stopping natural hemostatic drug for any bleeding and in compositions to care for gastritis and inflammation of the mucous membranes of the gastrointestinal tract (Duke 2002; Golovkin et al. 2001; Murav'eva et al. 2007; Yilmaz et al. 2019). Moreover, this species is widely used as a styptic for bleeding and hemorrhoids (Mashkovskii 2010). The active matter is thought to be azulenes, such as chamazulenes (Konovalov 1995). Also, the leaves and inflorescences contain sesquiterpene lactones, flavonoids, alkaloids stahidrine, achillein, and others. The content varies. Often, anthocyanins are present in the plant as polymorphous species (Ustyuzhanin et al. 1987), and the azulene/anthocyanin ratio changes among subspecies, showing a linkage in the accumulation of both compounds (Bashirova et al. 2003). Chamazulene, as the predominant drug, has mainly anti-inflammatory and antiallergic features (Jung et al. 1949). In flower brackets with red or rose petals, there is a greater yield of chamazulene (Ustyuzhanin et al. 1987). It is supposed that the main mechanism of its action consists in the inhibition of lipid peroxidation and participation in radical processes (Rekka et al. 1996, 2002), proazulenes and azulenes have padioprotectory activity (Roshchina et al. 1998b).

In Figure 3.1, analyzing secretory structures one can see the blue fluorescent images of biserial glandular trichomes (hairs) and glands on the petals of the flower as well as leaf glandular hairs. Biseriate (two row of cells) glandular trichomes (hairs) of the flower are shown in side view and top view and compared with pictures. The images may be observed with both luminescence and confocal microscopes under excitation by light at 360–380 nm. The fluorescence spectra of petal hairs registered by microspectrofluorimetry and laser-scanning confocal microscopy show maximum 490 nm in both cases and shoulder 510 nm in the first case. Earlier studies by microspectrofluorimetry, compiled in a monograph (Roshchina 2008), showed that floral glands also have more maxima—465, 475, 500, and 550 nm, depending on the sample and the stage of the plant's development. Figure 3.1 (images a–d) shows that when confocal microscopy was used, the main emission maximum of the petal

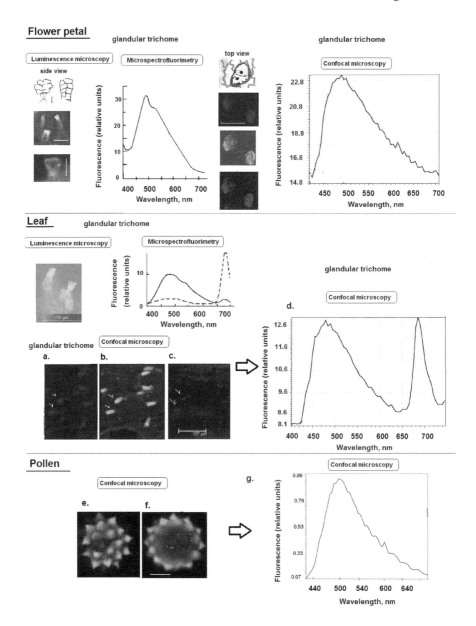

FIGURE 3.1 The fluorescence and emission spectra of glands and glandular hairs on the petal surface of flower, glandular hair on lower surface of leaf, and pollen from common yarrow, *Achillea millefolium*. Excitation 360–380 nm for luminescence microscopy and microspectrofluorimetry; laser 405 nm for confocal microscopy. Bars = 100 and 137 μm, respectively, for luminescence and confocal microscopy of glandular structures of petal and leaf or 25 μm for pollen. Confocal microscopy of leaf glandular trichomes has been demonstrated in blue (a), green (b), and red (c) channels, and for pollen (e, f) only in the blue channel. The emission spectrum from ROI is shown for secretory structures—leaf trichome (d) and middle of pollen (g). The dashed line for leaf hair is the emission spectrum for nonsecretory cells.

glands was 480–500 nm, and emission of leaf hair was shifted to a shorter wavelength (475–480 nm). Leaf glandular hairs contained phenols (maximum 480 nm) and chlorophyll (maximum 680 nm). Pollen also is a unicellular secretory system that fluoresces with maximum 480 nm according to microspectrofluorimetry (Roshchina 2008) and 490–500 nm as seen with confocal microscopy (Figure 3.1e, f).

Bright blue fluorescence due to sesquiterpene lactones and monoterpenes such as menthol predominated in the flower petals and leaves, which contain sesquiterpenoids (up to 1% of the essential oil, which includes about 30% azulenes, as well as the alkaloid achillein (Murav'eva et al. 2007). The fluorescence spectrum of crystalline azulene, which emits with maximum 430 nm, is shown in Figure 3.2. According to Murav'eva and coauthors (2007), secretory glands are fulfill of chamazulene (also fluorescing in blue at 420–435 nm), which may be transformed into the bitter prochamazulene achillicine. Diploid and tetraploid plants contain proazulene sesquiterpenes that are transformed into colored azulenes, including chamazulene (up to 25%) and achillicine. Amount of carotene also differs, but it weakly fluoresces upon contact with water (van Riel et al. 1983). Hexaploid plants are azulene sesquiterpene free. It should be noted that unlike leaf secretory cells, in petals, there is no chlorophyll in the glands, but some azulenes bound to cellulose in cell walls may also give red (maximum 620 nm) emission (Roshchina et al. 1995). Perhaps, some flavonoids (anthocyanins) and coumarins may contribute to the blue-green fluorescence (Roshchina 2003, 2008; Roshchina et al. 2011).

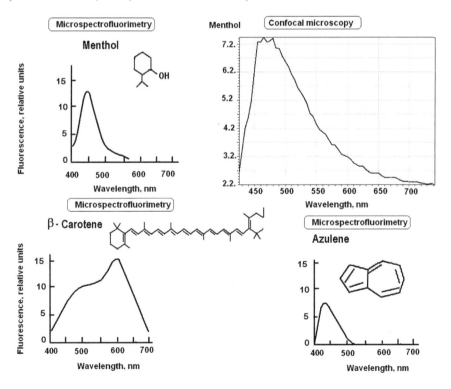

FIGURE 3.2   Main fluorescent compounds in Asteraceae family.

Microspectrofluorimetry has demonstrated that the composition of the compounds changes during the development of the secretory structures. In developing petal glands, sesquiterpene lactones such as azulenes predominate, as seen from the maximum 430 nm of the structures, while later, new compounds contribute to the glandular emission, when the emission maximum shifts to 480–500 nm (Roshchina 2008). Azulenes are also abundant in the developed leaves of white and red clovers, *Trifolium repens* and *T. pratense*, as well as in chloroplasts of leaves from *Pisum sativum* (Roshchina 1999, 2008; Roshchina et al. 2011a.).

Possible candidates to fluoresce at 400–530 nm in *A. millefolium* (Figure 3.2) are components such as the sesquiterpene chamazulene (maximum 430–460 nm in secretory structures), the proazulene austricine (maximum 430 nm), caffeic acid (maximum 450–460 nm), coumaric acid (maximum 520 nm) and flavonoid rutin (maxima of 470 and 550–560 nm). Figure 3.2 shows the main fluorescent components found in the family Asteraceae. *Achillea millefolium* also belongs to the taxon and is rich in azulenes, menthol and carotene.

The emission intensity per cell determined by microspectrofluorimetry and confocal microscopy can be found in the work of Roshchina et al. (2017b). In both cases, the secretory cells fluoresced with higher intensity in comparison with nonsecretory cells. The difference in emission varied from twofold to fourfold at 450–460 nm or 550 nm. Thus, the difference may be used for pharmaceutical practice in the estimation of the maturity of the medicinal plant as a whole and, perhaps, its enrichment in some fluorescent drugs in particular. In any case, the autofluorescence of secretory cells serves as a natural marker, and the fluorescence can be clearly seen among nonsecretory cells.

3.2 *ARTEMISIA ABSINTHIUM*, COMMON WORMWOOD

Common wormwood, *Artemisia absinthium*, known earlier as a classic stomachic stimulant of appetite, is today of medicinal interest as an anticancer drug (Wright and Colin 2002; Duke 2002; Efremov 2014). The herb contains 0.5–2% of essential oil (Murav'eva et al. 2007), which includes mono- and diterpenes that are weakly or non-fluorescent in the UV region, among them thujone, cadinene, phellandrene, and bisabolene. It may also contain the anticancer drug artemisinin, a sesquiterpene lactone (Zia and Chaudhary 2007). If the blue-green fluorescent surface images of the leaf glands and glandular hairs are compared with their emission spectra (Figure 3.3), one can see that secretions within the structures fluoresce with maximum 480 nm (spectra 1 and 2 for both glands and tips of glandular hairs) while weakly emitting in the middle of the hair with maximum 500 nm (spectrum 3). Chlorophyll, with maximum 680 nm, is seen weakly, if at all, by our optical probes.

Main fluorescent drugs may be azulenes, such as chamazulene and artemazulene, and proazulenes with the emission maxima in a range of 380–430 nm (Roshchina 2008). Bitter sesquiterpene lactone absinthin has maximum 550 nm (spectrum 4) measured from crystals of the substance by confocal microscopy. A similar maximum was found for ethanol solutions of absinthin. Artemisinin and o-coumaric acid emit with peak 500 nm (Roshchina et al. 2016b). The contribution of the main components absinthin and azulenes has not been seen, although the emission at 550–620 nm was observed (absinthin fluoresces with maximum 550 nm and would be better recorded if the excitation were 470 nm). In Figure 3.3b, fluorescing gland and the tip of glandular hair have the emission maximum 480 nm, similar to the maximum of o-coumaric acid. Azulenes in pure form emit in blue (see Figure 3.2), but it is known that they may be accumulated in cell walls (Roshchina et al. 1995), and in this case

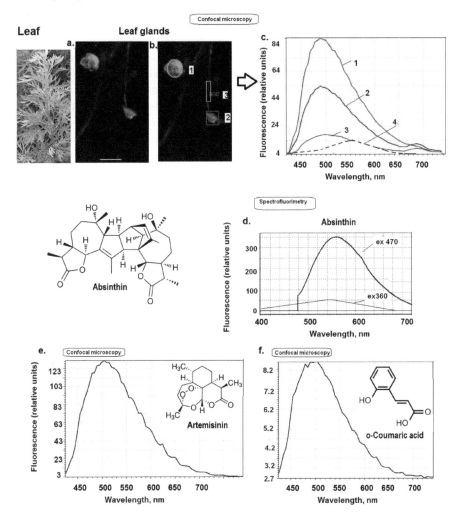

FIGURE 3.3 Images (a and b) and fluorescence spectra (c) of gland (spectrum 1), secretion in the tip of glandular hair (spectrum 2), and middle part of the hair (spectrum 3) on lower surface of leaf from common wormwood *Artemisia absinthium* (numbers of the measurable parts are in (a and b), bar=25 μm) and crystalline absinthin (spectrum 4), a characteristic compound for the species. Laser excitation 405 nm. The fluorescence spectra of the ethanol solution of absinthin (d) were recorded using a Perkin–Elmer spectropectrofluorimeter at the excitation 360–405 nm and 470 nm. The fluorescence spectra of pure individual drugs of the herb, crystalline artemisinin (e) and o-coumaric acid (f), were recorded by laser-scanning confocal microscope (laser excitation 405 nm).

after binding with cellulose the maximum of the compound was at 620–630 nm. We could see that secretions within the structures (Figure 3.3a, b) lightened with maximum 480 (spectrum 1 and 2 for both glands and tip of glandular hair), while weakly emitted middle of the hair – with maximum 500 nm (spectrum 3). On the comparing of the living cell emission with the fluorescence of individual components of *A. absinthium*, it should be remarked that chlorophyll maximum 680 nm practically has not been seen in the spectra in our samples (Figure 3.3c). The specific substance of the species is the bitter sesquiterpene lactone absinthin, from whose crystals emissions are measured by confocal microscopy, with peak 550 nm (spectrum 4 in Figure 3.3c). As seen in Figure 3.3, ethanol solutions of absinthin demonstrate a similar peak at an excitation of 360 nm measured by spectrofluorimetry, although more intensive emissions were observed at excitation by light at 470 nm (Roshchina et al. 2016b). Other individual components of *A. absinthium*, artemisinin and o-coumaric acid in crystalline form, emitted with peak 500–510 nm. Azulenes and proazulenes, emitting in blue, in cellular secretions may be masked by the above-mentioned components.

3.3 *ARTEMISIA VULGARIS,* COMMON MUGWORT

Common mugwort, *Artemisia vulgaris*, a well-known aromatic perennial plant from family Asteraceae with a characteristic scent, is used in officinal and folk medicine, mainly as an insecticidal and food plant (Wright 2002; Murav'eva et al. 2007; Anwar et al. 2015). All parts of the plant contain essential oils with all-purpose insecticidal properties (especially in the killing of insect larvae). Mugwort also has antibacterial and antiseptic characteristics; a paste of the leaves can be applied as a poultice for treating all kinds of skin infections. There is information about antispasmodic properties for menstrual pains and bronchodilatory properties for asthma. Mugwort stops diarrhea very effectively and may help in stomach ulcers, indigestion, and liver disorders; act as an emmenagogue, a nervine, a digestive, a diuretic, and a diaphoretic; and be used to flavor food. It has also been known as a home remedy for anxiety, restlessness, anxiety, insomnia, and depression. The essential oil and extract have appreciable antimicrobial potential. Nevertheless, one should be aware of the dangers. Mugwort used in large doses is toxic, especially for children and pregnant women, because it can cause uterine contractions. In some cases, pollens of the species are famous allergens and commonly cause allergies.

The mugwort plant contains essential oils (with cineole, and thujone), flavonoids, triterpenes, and coumarin derivatives, but thujone is toxic in large amounts or with prolonged intake (Murav'eva et al. 2007; Anwar et al. 2015). The composition of essential oils varies in different countries, and polymorphism takes place (Judzentiene and Budiene 2017). For Lithuanian species, the major constituents sabinene, 1,8-cineole, artemisia ketone, both thujone isomers, camphor, cis-chrysanthenyl acetate, davanone, and davanone B were found among 111 components. In the Turkish population from western Anatolia, the variation is larger, and there are clearer antimicrobial and oxidative effects of the essential oils (Erel et al. 2012). In folk medicine, tinctures of the species are used for the treatment of gastrointestinal diseases, mainly as appetite stimulants (Murav'eva et al. 2007). The leaves and buds, best picked shortly before mugwort flowers in July to September, are used as a bitter flavoring agent to season fat, meat, and fish. The active medicinal compounds of *Artemisia vulgaris* are flavonoids, coumarins, sesquiterpene lactones, volatile oils, inulin, and alkaloids (Anwar et al. 2015).

In our experiments, we saw blue or blue-green emission from leaves and flower petals when excited by UV or violet light, respectively (Figure 3.4). The fluorescence spectra were recorded by a microspectrofluorimeter, and maximum 460 nm with shoulder 550 nm (as well as the chlorophyll maximum 680 nm) was observed in the leaf secretory hairs, while nonsecretory cells had only one peak in red. In experiments with confocal microscopy, fluorescing glands and secretory hairs were seen in blue (weakly) and in green (brightly) channels when the 405 nm laser was used for the excitation (Figure 3.4a, b, d, e). The contribution of the red fluorescence specific to chlorophyll in the red channel (f), was also observed in the samples. In volatile oils, camphor, camphrene, germacrene D, 1–8-cineole, alpha-thujone, and β-caryophyllene are weakly fluorescent if at all. In leaf secretory cells (a–c), one can see maxima at 490–500 nm in glands (1, 2) and 450–460 nm in the hairs (3–5) as well as the chlorophyll peak 680 nm in all parts. In flower samples (d–h), secretory hairs, pollen, and pistils fluoresced, especially clearly in the green and red channels. The main maximum for all parts was 500–510 nm, while red fluorescence contributed with minimal intensity without a peak, which may be connected with the emission of flavonoids in the region. Keep in mind that most essential oils fluoresce in blue, unlike flavonoids, which emit in blue-green and/or green or even in red (Roshchina 2008). Phenols appear to mask the blue fluorescence of the oils because there is the combination of the compounds in oleophenol-containing secretory structures (Roshchina V.D. and Roshchina V.V. 1989; Roschina V.V. and Roschina V.D. 1993).

3.4   *BERBERIS VULGARIS* L., BARBERRY

Barberry or European barberry, *Berberis vulgaris* L., from the family Berberidaceae contains a large number of phytochemical materials. Extracts from the roots and leaves of *Berberis vulgaris* may have anticancer, anti-inflammatory, antioxidant, antidiabetic, antibacterial, analgesic, antinociceptive, and hepatoprotective effects (Golovkin et al. 2001; Duke 2002; Murav'eva et al. 2007) It includes more than 30 alkaloids, ascorbic acid, vitamin K, several triterpenoids, and about 10 phenolic compounds (Rahimi-Madiseh et al. 2017). Active matter of the plant is thought to be mainly alkaloids such as berberine, because this is the predominant active medicinal agent. It is possible to use different organs of *B. vulgaris*, especially fruit, to develop new drugs.

The task of analyzing pharmacological material seems to be more easily solved if the prevailing components are known. The extracts from the roots and, to a lesser degree, the leaves of the species and the individual alkaloid berberine itself may have anticancer, anti-inflammatory, antioxidant, antidiabetic, antibacterial, analgesic, antinociceptive, and hepatoprotective effects (Wink 2015; Rahimi-Madiseh et al. 2017). It is widely used against cholecystitis, chronic hepatitis, and infections in the post-puerperal period (Murav'eva et al. 2007). In traditional medicine, the fruit and flowers have not as yet been used for pharmacy. In view of this, we have studied

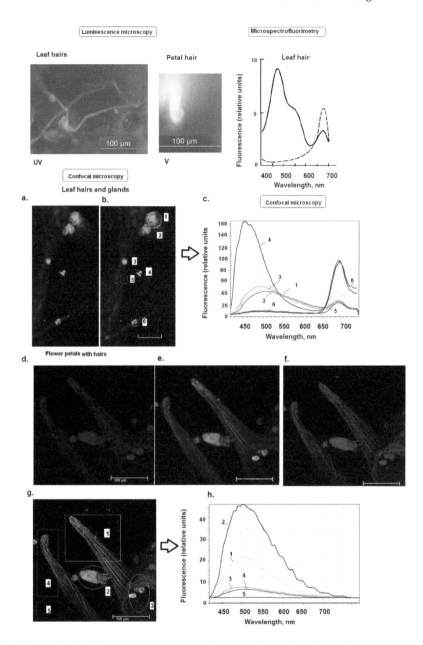

FIGURE 3.4 The fluorescence images and spectra of samples from *Artemisia vulgaris*. Luminescence microscopy of leaf and petal hairs – UV and violet; excitation by light at 360–380 nm or 350–430 nm. Microspectrofluorimetry of leaf hairs – the emission spectra of leaf secretory hair (solid line) and nonsecretory leaf cells (dashed line), excitation by light at 360–380 nm. Confocal microscopy of leaf and petal secretory cells – excitation by laser at 405 nm. (a and b) Images of leaf hair and glands or nonsecretory leaf cells; (c) the related numbers of ROI on image b; (d–g) flower petals with hairs; (d–f) the fluorescent images in blue, green, and red channels; (g) summa of optical slices; (h) the emission spectra, where numbers are related to image g.

these organs by various luminescence techniques—from the simple luminescence microscope and microspectrofluorimetry to confocal microscopy (Roshchina et al. 2017b). For the analysis of flowers of *B. vulgaris* as a potential pharmaceutical material, we also used stereomicroscopy of the tissue shown in Figure 3.5a and b. Microspectrofluorimetry recorded the fluorescence spectra of the main fluorescing parts of the flower—pistil and pollen (Figure 3.5c). Among the flower parts, the stigma of the pistil had maximum 540 nm, like berberine, while pollen had maximum 510 nm, which may show flavin characteristics (Roshchina 2008).

On excitation by the 405 nm laser under a confocal microscope, the surface of fruit containing secretory cells fluoresced in yellow (Figure 3.5d). In the fluorescence spectra, ROI for the cells (Figure 3.5e, spectrum 1) and the drop of secretion released out cells (Figure 3.5e, spectrum 2) demonstrated maximum 540 nm, similar to peak of pure berberine (see number 5 of the emission spectra in Figure 3.5f). In Figure 3.5 (image f), spectrum 3 shows the fluorescence of the fruit pulp (maximum 540 nm), and the hair emission ROI (spectrum 4) differs in the maxima position from berberine (perhaps due to phenols emitted in the range 470–500 nm). Earlier, a shift of the berberine emission in water solution (excited by light 488 nm) to the long-wavelength region occurred (Roshchina 2008). Unlike the surface secretory cells and secretion, the pulp of the fruit (Figure 3.5f, spectrum 3) had weak emission if any, while the secretory hair (Figure 3.5f, spectrum 4) seen on the surface, outside the fruit skin, emitted in blue. The secretory hair had maximum 460 nm (Figure 3.5f, spectrum 4), perhaps due to flavonols. Of course, we supposed that other components, such as the alkaloids columbanin and berbamine (Golovkin et al. 2001), can also contribute to the emission, but additional studies with these individual compounds are required.

3.5 *CALENDULA OFFICINALIS,* POT MARIGOLD

Pot marigold, *Calendula officinalis*, is used due to its antibacterial, antiviral, and anti-inflammatory properties (Murav'eva et al. 2007). Their flower baskets and pollen are rich in carotenoids (up to 3% of dry mass) and essential oils (Singh et al. 2011). In Figure 3.6, one can see the fluorescence image and spectra of the petal surface (a and b) and an individual glandular hair (d and e) from a sepal of the ligulate flower. The petal surface includes fluorescing secretory cavities or ducts, seen as the intercellular spaces (possibly filled with essential oils), and they emitted in blue with maximum 450 nm and small shoulder in a range shorter than 440 nm (Figure 3.6, spectra 1 and 2). In contrast, the main surface cells emitted weakly with maximum 550 nm (spectrum 3). This is similar to the known peaks for β-carotene crystals, which fluoresce with peaks at 520–534 nm or with shoulder 500 nm and peak 590 nm in concentrated ethanol solutions (Roshchina 2008). Unlike the base petal cells, the sepal glandular hair emitted in blue, green, and red. Under a confocal microscope, we could see the difference in the emission of capitate top cells of the intact multicellular hair (Figure 3.6, d and e). One of them fluoresced mainly in yellow (spectrum 1) and the second in the blue-green spectral region with a maximum of 475 nm (spectrum 2). In the cells of the stalk (just under tip of the hair), chloroplasts were seen, and chlorophyll brightly emitted in red with maximum 680 nm (spectrum 3). Pollen of the species

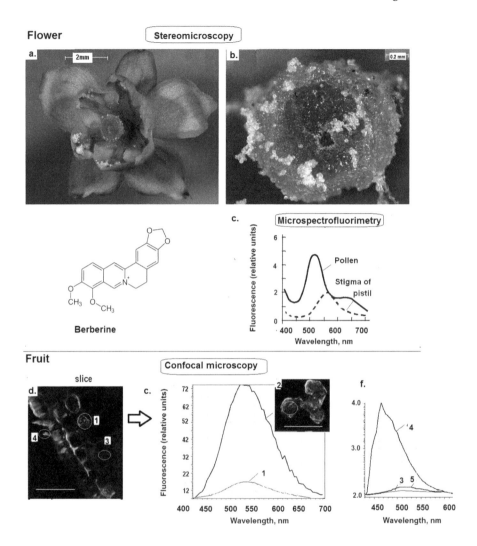

FIGURE 3.5 Fluorescence of alkaloid-containing secretory cells from *Berberis vulgaris* L. Flower: (a), (b) images of whole flower and pistil with pollen on the pistil stigma done by stereomicroscopy, (c) fluorescence spectra of pollen and pistil recorded by microspectrofluorimetry. Excitation 360–380 nm. Fruit cells fluorescing under Leica TCS SP-5 laser-scanning confocal microscope (laser excitation 405 nm). (d) Fluorescing cells of fruit border, transverse slice, bar = 300 μm; (e and f) the fluorescence spectra shown by ROI on (d) with numbers (1: cells emitting in green-yellow, 2: excretion from the fruit seen as on upper image, 3: pulp of fruit, 4: hair, 5: solution of 10^{-5} M berberine in water).

also fluoresces, mainly in green-orange (Figure 3.6, image g), and the middle part of the pollen grain has a maximum at 530–540 nm, similar to petals (Figure 3.6. image h, spectra 1–4). This peak is specific to β-carotene abundant in the orange petal (Roshchina 2008). Apertures, from which the pollen tube grows, emit more brightly with maximum 520 nm. The petals and pollen of *Calendula officinalis* contain up to

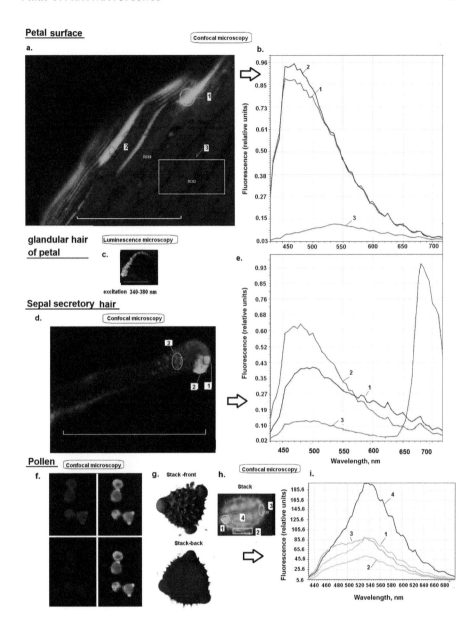

FIGURE 3.6 Laser-scanning confocal microscopy of the fluorescence images and spectra of *Calendula officinalis* (excitation by 405 nm laser). Petal of ligulate flower: (a and b) sum of fluorescing images and spectra of the surface, bar = 50 μm; (c) For a comparison, luminescene microscopy of biseriate hair of petal. Sepal secretory hair from the same flower: (d and e) sum of images and spectra in blue, green, and red channels, bar = 60 μm. Pollen: (f) images in three channels and sum; (g and h) sum of fluorescing images in stack; (i) spectra. Numbers on the images and spectra show concrete ROI (optical probe).

3% of dry mass of carotenoids such as β-carotene and its oxidized derivatives vio-loxanthin, flavoxanthin, auroxanthin, and so on (Murav'eva et al. 2007), which may contribute to the fluorescence. Bright green emission may also be due to β-carotene (which is abundant in the orange petals).

3.6 *CHELIDONIUM MAJUS,* GREATER CELANDINE

An example of alkaloid-rich species is greater celandine, *Chelidonium majus* L. (fam. Papaveraceae), which has segmented laticifers filled with yellowish-brown contents. Medical application of the alkaloids is known in practice for cauterization of warts and condylomata and in the treatment of papillomatosis (Murav'eva et al. 2007; Efremov 2014). Furthermore, the herb has many other features—spasmolytic, cholagogue, bitter, alterative, diuretic, laxative, anodyne, purgative, caustic, antiallergic, anti-inflammatory, and abortifacient—and is applied internally for jaundice, gallstones, and gallbladder disease or externally for eczema, verrucas, and warts as well as in antitumor therapy (Murav'eva et al. 2007; Aniszewski 2007). In the herbaceous parts and roots of the species, alkaloids predominate—up to 2% and 4%, respectively. There are about 20 alka-loids, including chelerythrine, sanguinarine, and berberine. The brownish color of the secretion may relate to the presence of berberine (Murav'eva et al. 2007). Extracts from the herb possessed marked cytotoxicity, suppressing the growth of cultivated human lymphoblastoid Raji cells (Spiridonov et al. 2005). Perhaps this was due to the effect of the alkaloid sanguirythrine, which acts in the same way in pure form at concentrations of 1–2 mg/mL.

Figure 3.7 (images a–d) with color photos represents the native fluorescence of a longitudinal slice of greater celandine *Chelidonium majus* stem, seen under a luminescence microscope at excitation by 400–430 nm light. This is orange emis-sion within the stem interior (image a) or greenish-yellow in the flower petal with blue-lightening laticifers (image b). Under a laser-scanning confocal microscope in appropriate pseudocolors the laticifers emitted in yellow on the red-fluorescing leaf surface (image c) or as an optical slice through the leaf tissue (image d) abundant in yellow laticifers. The optical slice of the leaf reveals yellow-orange laticifers parallel to the veins.

According to data received from microspectrofluorimetry and confocal micros-copy in Figure 3.7, the emission spectra of laticifers in flower and stem demonstrated maxima at 595 nm, while outside the secretory cells, the peak shifted to a shorter wavelength range, as noted earlier (Roshchina 2008). Regarding laticifers of the leaf analyzed by confocal microscopy, one can see the fluorescence spectra of laticifers with a characteristic main maximum 540 nm and other peaks at 550, 570, and 590 nm as well as a small peak at 680 nm due to chlorophyll if the object was excited by the 405 nm laser. In leaf parenchyma, there are no laticifers, and chlorophyll with maximum 680 nm predominated (Roshchina et al. 2016b). The comparison of the latex emission with the fluorescence spectra of pure chelerythrine and sanguinarine, both compounds having equal maxima at 590 nm, demonstrated the complex nature of the secretory composition of the structure (the dashed line in Figure 3.7f shows the spectrum of chelerythrine).

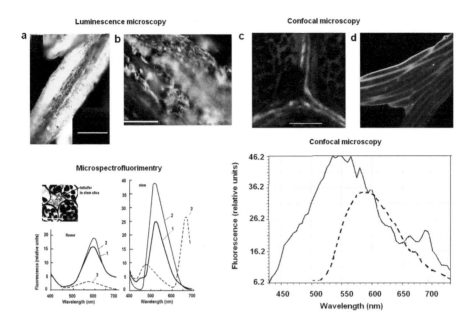

FIGURE 3.7   Fluorescent images and fluorescence spectra of alkaloid-containing secretory cells of greater celandine, *Chelidonium majus.* (a) Image of longitudinal stem slice, excitation 400–430 nm, observed under Leica D 6000B luminescence microscope. The brown or yellow-orange laticifers are seen; bar = 200 μm; (b) part of the flower petals with blue-emitting laticifers, bar = 100 μm; (c) and (d) leaf images under confocal microscope, excitation laser 405 nm. (c) Part of external surface of leaf with yellow-fluorescing laticifers (one channel, bar = 200 μm); (d) optical slice (in three channels—blue, green, and red); bright yellow-orange fluorescing laticifers along the veins are seen to stand out against the red background of chlorophyll-containing cells. Bar = 80 μm. (e) Microspectrofluorimetry: secretory cells as leaf and stem laticifers (stem laticifer (1), latex (2), and nonsecretory cells (3)). (f) Confocal microscopy: leaf laticifer (solid line) and crystal of the alkaloid chelerythrine (dashed line).

If the fluorescence spectra of living cells are compared with possible drugs present in the common celandine (Figure 3.7), a researcher may see that unlike leaf laticifers, the laticifer of flower petals and its latex had a maximum at 580–590 nm, corresponding to the maximum of alkaloids berberine, chelerythrine, and sanguinarine (Figure 3.8). Póczi and Böddi (2010) studied laticifers and latex of greater celandine laticifers and the native spectral properties of the latex in various organs. Whole plants have been analyzed in a gel documentation system using an ultraviolet light source, while the localization of the laticifers was observed along the leaf veins under a fluorescence microscope using blue excitation light. When different tissue pieces were measured, fluorescence spectroscopic studies showed that the greater celandine alkaloids have emission wavelengths at 469, 530–531, 553, 572–575, and 592 nm and excitation bands from 365 to 400 nm. Direct measurements on tissue pieces (without the extraction and the separation of the components) provided information about the complexity of the latex and the various ratios of the alkaloid contents in the tissues. These results may allow conclusions to be made about the

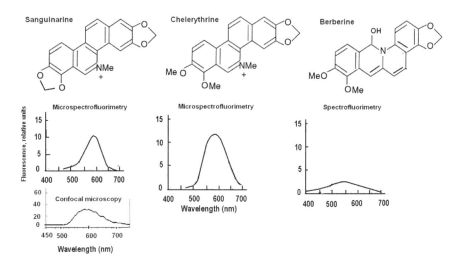

FIGURE 3.8   Fluorescence spectra of alkaloids contained in leaf and stem secretory cells of greater celandine *Chelidonium majus*, recorded by microspectrofluorimeter MSF-1, confocal microscope Leica TCS SP-5, laser 405 nm excitation, and spectrofluorimeter Perkin–Elmer 350 MPF 44B.

alkaloid contents and composition or the ratios of the alkaloid components in various species of celandine.

3.7   *CICHORIUM INTYBUS,* CHICORY

Chicory, *Cichorium intybus* (family Asteraceae), has been used as a medicinal and food (a coffee substitute) plant since ancient times in Europe and Asia. Its root useful features are covered in reviews (Nandagopal and RanjithaKumari 2007; Chandra and Jain 2016), and the species is recommended for treating different ailments from wounds to diabetes; antibacterial effects of the plant on *Bacillus subtilis*, *Staphylococcus aureus*, and *Salmonella typhi* have also been described. The whole plant contains a number of medicinally important compounds, such as inulin, esculin, volatile compounds (monoterpenes and sesquiterpenes), coumarins, flavonoids, and vitamins. It possesses hepatoprotective, gastroprotective, cardiovascular, antioxidant, hypolipidemic, anticancer, reproductive, antidiabetic, anti-inflammatory, analgesic, sedative, immunological, antimicrobial, anthelmintic, anti-protozoal, wound-healing, and many other pharmacological effects (Al Snafi 2016) due to its content of sesquiterpene lactones (especially lactucin, lactucopicrin, 8-desoxy lactucin, and guaianolid glycosides, including chicoroisides Band C and sonchuside C) and caffeic acid derivatives (chiroric acid, chlorogenic acid, isochlorogenic acid, dicaffeoyl tartaric acid, etc.). The flower and leaf are also used in folk medicine to treat liver disorders and inflammation, because their extracts include cyanidin-3-O-(6″-malonyl-β-glucopyranoside) as the major anthocyanin (>95%), as described by some authors (Mulabagal et al. 2009).

Pharmacological raw materials from *Cichorium intybus* contain many secretory cells, such as glands in flower petals, pistils, and pollens (which are also secretory single cells). The luminescence microscopy after excitation by UV or violet light (Figure 3.9) showed that blue petals rich in anthocyanins weakly fluoresced (image a). However, their glands demonstrated bright blue or yellow emission, (images b or c). Yellow fluorescence is better seen after excitation at 450–490 nm. Microspectrofluorimetry of the emission intensity also demonstrated maximum at 680 nm (chlorophyll). Extracts from petals (w/v 1: 1) excited by light at 360 nm fluoresced in different regions of the spectrum depending on the solvent and differential

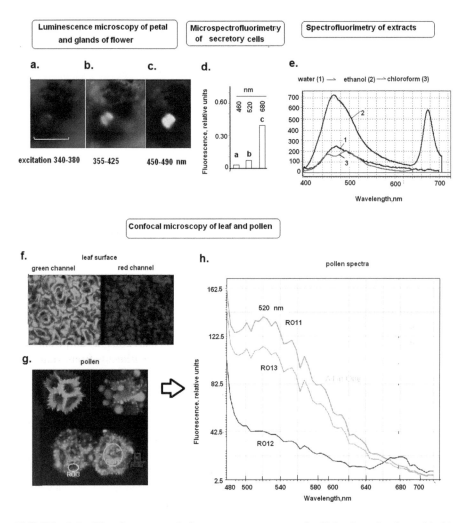

FIGURE 3.9 The images and fluorescence spectra of *Cichorium intybus*. (a)–(c) Luminescence microscopy images of flower petals (upper side) and isolated glands (lower side) after excitation by light at 360–380 nm (a), 430 nm (b), and 450–490 nm (c). Bar = 100 µm. (d) Microspectrofluorimetry of the glands' emission intensity after excitation by light at 360–380 nm (e) Spectrofluorimetry of petal extracts (1: water, 2: ethanol, 3: chloroform) after excitation by light at 360 nm. (f) Confocal microscopy of leaf surface and pollen after excitation by laser at 488 nm.

extraction as recorded by spectrofluorimetry. Water extracts from petals and leaves emitted only in blue-green with maximum 470 nm, while ethanolic extracts had two maxima 470 and 675 nm (the latter peak is specific to chlorophyll). A final extraction with chloroform showed fluorescence maxima 450 (which appears to be due to terpenoids) and 490 nm. Confocal microscopy of leaf surface in green and red channels did not show visible glands, but demonstrated the complex fluorescing structure of pollen.

Unlike petals and leaves, pollen fluoresced in green under a confocal microscope, and the application (combination) of summa of optical slices collected together demonstrated the complex structure of the male microspores with their reproductive function. Drops of secretion penetrating through the apertures of the pollen grain can be observed. The ROI (region of interest) on the pollen showed a main maximum 520 nm in the fluorescence spectrum, both for the center of the pollen grain (ROI 1) and for the aperture part (ROI 2); sometimes, parts of immature pollen can include a small content of chlorophyll (ROI 3, violet color), which are not present in mature pollen.

3.8   *EQUISETUM ARVENSE,* HORSETAIL

Horsetail, *Equisetum arvense*, is a widely used medicinal herb (Golovkin et al. 2001; Duke 2002; Murav'eva et al. 2007; Asgarpanah and Roohi 2012). Extracts of this species possess various pharmacological features (Sandhu et al. 2010)and have been used for urinary diseases as diuretics (Syrchina et al. 1974; Murav'eva et al. 2007), for osteoporosis (Radole and Kotwal 2014), for pleuritis, and as hemostatic (styptic) drugs for various hemorrhages (Murav'eva et al. 2007). Moreover, hydroalcoholic extracts have demonstrated sedative and anticonvulsant effects (Dos Santos et al. 2005). The species possesses antioxidant, anticancer (Al Mohammedi et al. 2017), antimicrobial, smooth muscle relaxant (effects on the vessels and ileum), anticonvulsant, sedative, anti-anxiety, dermatological, immunological, antinociceptive, antiinflammatory, antidiabetic, diuretic, inhibition of platelet aggregation, promotion of osteoblastic response, anti-leishmanial, and many other effects, according to a review (Al-Snafi 2017).

The ability to autofluoresce is specific to various terpenoid and flavonoid compounds. Twenty-five compounds were identified in essential oil of *Equisetum arvense*, of which hexahydrofarnesyl acetone (18.34%), cis-geranyl acetone (13.74%), thymol (12.09%), and trans-phytol (10.06%) were the major constituents (Radulović et al. 2006), and very strong antimicrobial activity against all tested strains of bacteria and fungi was observed. The main components also include 3,7,11,15-tetramethyl-2-hexadecen-1-ol, hexadecan-1-ol, esters of linolenic and octadecenoic acids, as well as phthalic and gibberellic acids (Yilmaz et al. 2014). Phytochemical analysis detected the presence of tannins, saponins, sterols, and flavonoids. Up to 5% of the triterpene saponin equisetonin is present in the herb, while small amounts of the alkaloids nicotine, palustrine, and 3-methoxypyridine are biologically active (Murav'eva et al. 2007; Yilmaz et al. 2019). Possible active components include the flavonoid equisetrin, salicylic acid, and silicic (silicon) acid (up to 25% bound to organic compounds). But horsetail cells, usually rich in silica deposits, fluoresce only after treatment with special dyes (Law and Exley 2011. Flavones are abundant in the horsetail herb, such as apigenin (Syrchina et al. 1974), genquanin, and luteolin (Syrchina et al. 1978). Isoquercitrin is the main flavonoid, and a considerable amount of di-E-caffeoyl-meso-tartaric acid has been found in extracts with various alcohols. Water extracts contain a high level of phenolic acids and a low percentage of flavonoids (Mimica-Dukic et al. 2008). The large amount of silica in the plant is valuable for collagen metabolism (Radole and Kotwal 2014). Extracts have significant peroxyl radical scavenging activity (Četojević-Simin et al. 2010). In fertile sprouts of *E. arvense*, some authors (Jun, Li-Jiang and Ya-Ming 2001) found three new phenolic glycosides.

The horsetail herb fluoresces in different ranges of the spectrum depending on the stage of development, as seen in Figure 3.10. Young plants (images a and b) have active secretory structures termed *hydathodes*, which release surplus water and solvated compounds (image b), emitted under 360–380 nm UV light at 490–500 nm (secretion) or 520–540 nm (hydathode), which may be related to flavonoids or flavins respectively (Roshchina 2008). The leaf endings of horsetail are characterized by the free entrance of fluorescent central vein (image c), which on excitation by UV light at 360–380 nm, emits in blue-green, with one maximum 520 nm in the spectrum. This fluorescence is peculiar to flavonoids (480–490 nm) and flavins (500–520 nm) (Roshchina 2008). Apigenin isolated from the herb shows bright blue fluorescence in UV light (Syrchina et al. 1974). The vein emission differed from that of mesophyll cells rich in chlorophyll (a large maximum at 675–680 nm). On excitation by blue (450–490 nm) light, the vein is seen in yellow (image c). The adult species is usually analyzed in pharmacy, although there is a sporophyte stage (images e and f) with the formation of sporangia with microspores (image g). The microspores, with a double set of chromosomes, are termed *vegetative*, unlike the male and female spores that develop later. The vegetative microspores (image h) fluoresce in blue-green due to cell wall phenol inclusions (Roshchina 2007a, b) and in red, showing the presence of chlorophyll. The fluorescence spectra (image i) reflect emission from the spores surrounded by elaters (threads serving for attachment to the soil), as shown in spectrum 1 with maxima 460 (flavonoids), 550 (carotenoids, alkaloids), and 680 nm (chlorophyll). Elaters lack chlorophyll (spectrum 2). When a microspore loses its elaters (spectrum 3), it has maxima 480 (flavonoids), 550 (carotenoids, alkaloids), and 680 nm (chlorophyll). Laser-scanning confocal microscopy reveals the emissions from the surface (image j) and after optical slicing (image k) respectively. The surface green fluorescence is clearly visible, and the slice shows chlorophyll in chloroplasts within the microspores. Extracts of spore-bearing stems of *Equisetum arvense* L. (field horsetail) contain saponaretin, apigenin 5-glucoside, luteolin 5-glucoside, kaempferol 3-sophoroside, quercetin 3-glucoside, and 4-hydroxy-6-(2-hydroxyeth yl)-2,2,5,7-tetramethylindanone, a compound of ketonic nature. This last substance was isolated previously from an extract of the herbage of the yellow field horsetail and identified by high-performance liquid chromatography (HPLC) (Syrchina et al. 1980). It appears that the surface fluorescence occurs due to phenol compounds. Methanolic extracts from the overground sporophytes of all species of the subgenus *Equisetum* (*E. arvense* L., *E. bogotense* HBK, *E. fluviatile* L., *E. palustre* L., *E. pratense* L., *E. sylvaticum* L., and *E. telmateia* Erh.) have been studied by HPLC. Interspecific and intraspecific variation of phenols, mainly flavonoid glycosides, was found in these species (Veit et al. 1995).

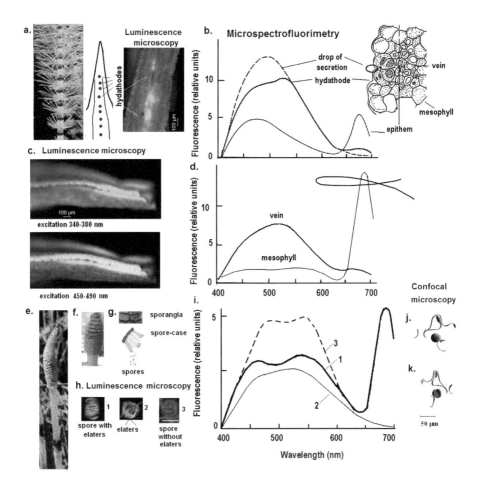

FIGURE 3.10 The images, fluorescence, and fluorescence spectra of various parts of horse-tail, *Equisetum arvense*, recorded by luminescence microscope Leica DM 6000 B (a, c, h), microspectrofluorimeter MSF-1 (b, d, i), and laser-scanning confocal microscope LSM 510 NLO Carl Zeiss (j, k). (a) Young plant with active hydathodes that contain fluorescent secre-tion (excitation at 340–380 nm); (b) fluorescence spectra of hydathode; (c) leaf ending with fluorescent vein that releases secretions; (d) fluorescence spectra of leaf ending; (e and f) sporophyte with sporangia; (g) scheme of individual sporangium full of vegetative micro-spores; (h) microspores (1: common view, 2: elaters, 3:cell without elaters) fluorescing under Leica DB 6000 luminescence microscope (bar = 30 μm); (i) fluorescence spectra of the micro-spores; (j) and (k) views of the microspores under laser-scanning confocal microscope from the surface and in an optical slice, respectively.

All species have shown quantitative and qualitative variations during plant devel-opment. As an example, the authors (Veit et al. 1993) present chromatograms of dif-ferent developmental stages of *E. arvense* and the accumulation dynamics of some phenols in that species. Phenols accumulating in different organs of the plant are distinct. Apart from the spores, only diploid tissues are able to accumulate flavo-noids. The haploid vegetative gametophytes accumulate large amounts of caffeic acid esters and styrylpyrones. The rhizomes are also free of flavonoids but contain various styrylpyrones (Veit et al. 1993) as well as hydroxycinnamoyl esters. Equisetumpyrone occurs in *E. arvense* vegetative gametophytes but fluoresces only in the UV spectral region (Veit et al. 1993), unlike flavonoids. The intense fluorescence of vegetative microspores may be due to emission by flavonoids. Styrylpyrones seem to replace flavonoids in the gametophytes and sporophytic rhizomes of *Equisetum* species. We currently have no data about the pharmacological importance of the vegetative microspores, although they actively germinate and contain many valuable flavonoids. All of these, and aromatic acids, may fluoresce both in the whole sporophyte and in microspores. Vegetative microspores are a potential pharmaceutical material, like *Lycopodium* spores: naturally manufactured, super-robust biomaterial for drug delivery (Mundargi et al. 2015).

3.9 *FRANGULA ALNUS,* BUCKTHORN

Bark particles from buckthorn, *Frangula alnus* Mill. or *Rhamnus frangula* Mill. (family Rhamnaceae), are known as laxative or bowel tonic relaxants. Red-colored anthraquinones are anthracenic derivatives that are found in some medicinal plant species (Golovkin et al. 2001; Duke 2002; Murav'eva et al. 2007; Efremov 2014). They are rich in oxymethyl anthraquinones such as frangula-emodine and frangulin in A and B forms with different residues, for example, gluco-frangulins. The dried bark contains up to 8.5–9.1% of the anthracene derivative as determined at absorbance wavelength 524 nm (Kurkin et al. 2014). In our work, the raw dried material showed red emission in the main tissue under a luminescence microscope (Figure 3.11a). Optical probes by confocal microscopy (images b) in three different channels—blue, green, and red—showed the different locations of fluorescing compounds. The complex spectra from different parts (ROI) of the bark cells excited by the 405 nm laser have maxima in the range 490–500 nm in blue and blue-green (spectra 2–4) and mainly (spectra 5 and 6) in orange-red with maxima 590, 600, 620, 640, and 680 nm. The presence of chlorophyll (clearly visible in the red channel of the confocal microscope in Figure 3.11 [spectrum 6]) can also be seen. The maximum 680 nm is especially intense when the sample is excited by red laser light of 680 nm.

The most intense emission under the laser-scanning confocal microscope in three channels—blue, green, and red—is seen after laser excitation of 405 nm (Figure 3.11b). Anthracene derivatives, such as frangulin, are crystalline substances of yellow or orange-red color and fluoresce in red when exposed to UV light. Confocal microscopy identifies a complex spectrum with maxima in blue, green, and red. The blue channel demonstrates emission with maxima of 500–520 nm in different ROI (images 2–4). In the red channel, we can see red emission with maximum 600–620 nm (image 5) specific to frangulin. The presence of chlorophyll is also seen in the parts excited by red laser light of 561 nm (image 6).

A comparison with the fluorescence of water and ethanol extracts (Figure 3.12) demonstrated similarity with the emission from native cells, unlike water extract (which fluoresces with maximum 510–520 nm), while red pigments fluoresce at 600–650 nm.

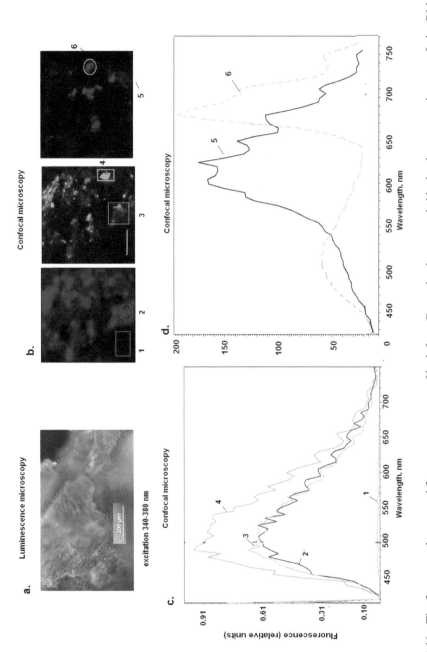

FIGURE 3.11 The fluorescent images and fluorescence spectra of bark from *Frangula alnus*, recorded by luminescence microscope Leica DM 6000 B (a) and laser-scanning confocal microscopy Leica TCS SP-5, with 405 nm laser, in three channels (b). ROI are shown as numbers on the images (b) and spectra (c and d) with excitation by laser 405 nm (spectra 1–5) and 561 nm (spectrum 6).

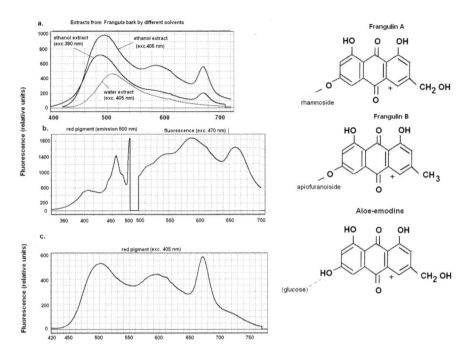

FIGURE 3.12 The fluorescence spectra of extracts and purified red pigment (possible formulae given on the right) from bark of *Frangula alnus*, recorded by spectrofluorimeter Perkin–Elmer 350 MPF-44B (a and b) and by laser-scanning confocal microscopy Leica TCS SP-5, with 405 nm laser (c).

3.10 *HUMULUS LUPULUS* L., COMMON HOP

Hop, *Humulus lupulus* L., is used as a natural sedative drug to calm the nerves and reduce stress and anxiety (Duke 2002; Chadwick et al. 2006; Murav'eva et al. 2007; Efremov 2014). It has anticancer effects (Fang and Ng 2013) and demonstrates estrogenic properties (Chadwick et al. 2006). For centuries, this plant has been used to reduce nervous tension or to promote a good night's rest by counteracting the effects of sleeplessness and insomnia. Moreover, it is also well known as the preservative and flavor used to make beer. There is a good publication about secretory structures (Melnychuk et al. 2013), in which channels of the schizogenous-lysigenous type, active during the stage of rapid shoot growth, have been discovered. Most information about the cultivation and use of hops is concentrated in a special atlas devoted to the species (Melnichuk et al. 2014).

The main active matter are bitter acids humulone and lupulone, as well as weak acid components. In hop pharmaceutical raw material, there are the bract multicellular glands of matured female flowers ("female cones") containing phytoestrogens (Chadwick et al. 2006). A phytoestrogen, 8-prenylnaringenin, has recently been discovered in the female flowers of hops (Chadwick et al. 2006). In many cases, extracts of *H. lupulus* cones may also act as pro-oxidants due to the presence of flavonoids (Olšovská et al. 2016; Karabin et al. 2016) and have antitumor potential (Fang and Ng 2013;

Karabin et al. 2016). All active hop components accumulate in the multicellular glands (our stereomicroscopic images on Figure 3.13a–c), which when mature, release small glandular cells named *lupulin* in pharmacy (Figure 3.13b and c). The nuclei of the single glandular cells were visualized after staining by Hoechst 33342 fluorescent dye for DNA (see Chapter 4). For therapy, the dried lupulin powder is used. We had no previous information about the autofluorescence of lupulin. Only MacWilliam (1961) had seen blue and yellow fluorescence excited by UV light on paper electrophoresis of hop resin in zones corresponding to humulone and lupulone. In our experiments, one can see blue-green fluorescence of the multicellular gland under UV excitation in confocal microscope and emission spectra with maximum 510–520 nm were recorded by microspectrofluorimetry (Figure 3.13d and e). The bitter acids humulone and lupulone, separated on silica gel chromatographic plates in UV light, emitted in blue (Wagner and Bladt 1996). In our experiments, bitter acid–enriched water extract from hop glands (0.5 mg/5 mL) emitted with maximum 475–480 nm (Roshchina et al. 2017b). Bitter acids, flavonoids (which fluoresce at 460–480 nm), and phenolestrogen may contribute to the emission.

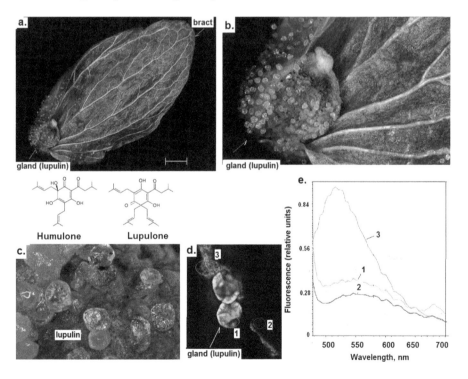

FIGURE 3.13 Fluorescence of pharmaceutically valuable secretory cells of female flower (cone bracts) from *Humulus lupulus* L. (a)–(c) Stereomicroscopic images of multicellular resinous glands, from which single cells are liberated and named lupulin bar = 1 mm; (d and e) common image of glandular surface fluorescing under Leica TCS SP-5 laser-scanning confocal microscope (laser excitation 405 nm) in green-yellow and spectra. ROI of the image (marked by blue and violet rings, numbers 1, 2, 3.

3.11 *HYPERICUM PERFORATUM,* COMMON ST. JOHN'S WORT

Common St. John's wort, *Hypericum perforatum*, is known as a medicinal plant from the family Hypericaceae, which has been used as an anti-inflammatory, binder/astringent, and antiseptic for treating diarrhea and colitis as well as for rinsing and lubricating gums in stomatology (Newall et al. 2002; Duke 2002; Ernst 2003; Rao et al. 2005; Murav'eva et al. 2007). It may be used as an antidepressant (Chatterjee et al. 1998). A broad spectrum of biological activities, including against AIDS, was declared by Lavie and coauthors (1995), who described the main mechanism of cellular damage due to activation by light that led to the production of free oxygen or/and semiquinone-type radicals responsible for the virucidal effects. Scientific medicine recommends St. John's wort preparations both externally and internally for soothing. Folk medicine uses the red, oily extract of *H. perforatum* to treat and heal wounds (Samadi et al. 2010; Süntar et al. 2010). St. John's wort oil is a radical treatment for burns over a large area of tissue destruction, even if two-thirds of the body surface is affected. Standardized preparations are commercially available, and most research has studied alcoholic extracts and isolated compounds. Two main compounds are found in *Hypericum* species, considered as active materials with antibiotic properties (Liu et al. 2005): the red-colored anthraquinone hypericin (Brockmann et al. 1950) and the colorless prenylated phloroglucinol hyperforin (Bystrov et al. 1975, 1976), as well as some flavonoids. Their properties have been studied for application to medicine (Schulz et al. 2005). The main pigment of oil ducts, colored red, is used as a medical drug (Kariotti and Bilia 2010). According to Vattikuti and Ciddi (2005), the herb contains (%) up to: pseudohypericin (5–6); protohypericin (7–8); phloroglucinol derivatives such as hyperforin and furohyperforin (9); adhyperforin (10); xanthones (11); essential oil (12); α- and β-monoterpenes, pinenes, limonene, and sesquiterpenes (13); flavonols: catechins (14); flavonoids, including hyperoside (15); quercetin and quercetrin (17); rutin (16); biapigenin (18); and kaempferol (19–20). Red pigments related to hypericins accumulate in oil glands, pistils, and fruits, probably as a plant defensive compound.

Conventional microscopy is useful when the samples of *Hypericum perforatum* contain pigmented secretions that are readily visible, as can be seen in Figure 3.14a, b, c, g, h, in particular for the leaf oil ducts (g) and secretory glands of both leaf (b) and flower petals (h), where the red pigment hypericin is easily visible. Also, on the leaves there are colorless secretory glands that can be seen visually and under transmitted light in both conventional and confocal microscopes (Figure 3.14c–f). The absorbance spectrum of hypericin has maxima 590 and 595 nm (Roshchina 2008). Colorless secretory cells do not contain the pigment. The fluorescence of hypericin components on excitation at wavelength 315 nm was in the region of 590 nm, while hyperforin emitted in the UV range (Bauer et al. 2001). In our experiments by luminescence microscopy, including confocal microscopy, one can see fluorescence of glandular cells excited by UV (360–380 nm) or violet light (355–425 nm) (Figure 3.14, image k). Confocal microscopy (Figure 3.14) enables the pigment fluorescence to be seen in leaf (images and spectra l–o) anther and pollen (image p). An ethanol solution of hypericin extracted from the surface of petals (for 20 min) fluoresces with maximum 608 and shoulder 650 nm when excited by light at

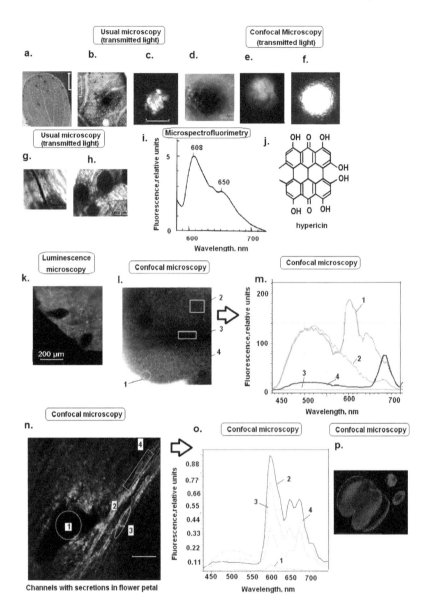

FIGURE 3.14 Leaf oil ducts and secretory cells of *Hypericum perforatum*. (a) Leaf with dark and light spots of secretory structures, bar = 500 mm; (b), (g), (h) secretory structures with red pigment hypericin seen in transmitted light of conventional microscope. Bars = 500, 200, and 100 μm; (c)–(f) colorless gland seen in transmitted light of conventional microscope or confocal microscope, bars = 200 and 100 μm; (k) petal glands with red pigment under luminescence microscope Leica DM 6000 B at excitation 355–425 nm, bar = 100 μm; (l–o) fluorescence images and spectra with ROI of leaf and flower petal; or (p) image of anther and pollen under confocal microscopes Leica TCS SP-5, laser 405 nm, or Zeiss 510 NLO, laser 458 nm, bars = 100 or 20 μm; (i) the fluorescence spectrum of ethanolic solution of hypericin, (j) excitation 360 nm by spectrofluorimeter Perkin-Elmer 350.

360 nm (Figure 3.14i, j). Comparing the emission spectra of isolated hypericin with the living cell, one can see the similarity in the position of maxima in the yellow-red region. The fluorescence spectra of leaf gland and channels-ducts (Figure 3.14m and o) clearly show peaks related to hypericin at 600 nm (ROI spectra 2 or 3 for images l and o), while parts outside the secretory system show only maximum 675–680 nm (ROI spectra 4 for images m and o), specific to chlorophyll. It should be noted that in leaf mesophyll and gland, phenols (flavonoids) may fluoresce in the 450–550 nm region with maximum at 500–520 nm. Hyperforin also has antibacterial effects (Schempp et al. 1999).

3.12 *LEDUM PALUSTRE*, LABRADOR TEA, CRYSTAL TEA LEDUM, MARSH TEA

Labrador tea or marsh tea, *Ledum palustre* L., grown in peaty soils, shrubby areas, and moss and lichen tundra, is a species of the family Ericaceae that is used in both officinal and folk medicine (Murav'eva et al. 2007; Zhang et al. 2017). The raw pharmaceutical material contains a mixture of cut stems, leaves, and a few fruits. The whole plant is rich in green essential oil (up to 7%) with a sharp smell that causes a headache in some people. It is also used in homoeopathy for the treatment of rheumatism, arthrosis, and insect bites. The expectorant and antitussive effect of marsh tea is due to the ledol contained in the plant's essential oil. Only one herbal product, called "Ledin," is produced from this plant material; it is based on essential oils and used as an antitussive. Currently, drugstores are stocking the packed raw material, which is very promising for the development of new medicines (Upyr et al. 2015). The essential oil of *Ledum* is rich in sesquiterpene lactones (Zhang et al. 2017). The biological activity of the Labrador tea volatile constituents in secretions is so high that they may be used for chemical degradation of plastic (Xiu et al. 2003).

The autofluorescence of Labrador tea appears exclusively in blue from both cells and secretory cavities (Roshchina et al. 2017b), as seen in Figure 3.15. In stems, the sesquiterpene alcohol ledol (a pharmaceutical substance) predominates in the fluorescence under UV light in a luminescence microscope at different magnifications (image a). On excitation by longer-wavelength light there is no fluorescence except for the chlorophyll-containing parts, which emit in red due to chloroplasts. Laser-scanning confocal microscopy demonstrated the difference between blue-green (b [ROI] and c [spectra 3–5]) in the range 450–620 nm and red fluorescing (b [ROI] and c [spectrum 2], due to chlorophyll) cells in the leaf fragment (upper side). Another sample of this object (lower side) shows a small peak for chlorophyll and the highest maximum at 460 nm (d and e [spectra 2 and 4]).

The blue range includes fluorescence from several compounds: sesquiterpenes emitting with maxima at 430–450 nm and flavonoids that may fluoresce at 460–480 and 500–550 nm. Only spectrum 4 has no contribution from chlorophyll (maximum 680 nm). Blue fluorescence at 430–450 nm is usually related to ledol (Figure 3.15, spectrum f), although in this sample, the blue pigment azulene also was found by thin layer-chromatography (the absorbance spectrum g). As early as 1926, the greenish or reddish color of essential oil from Labrador tea *Ledum grenlandicum* and

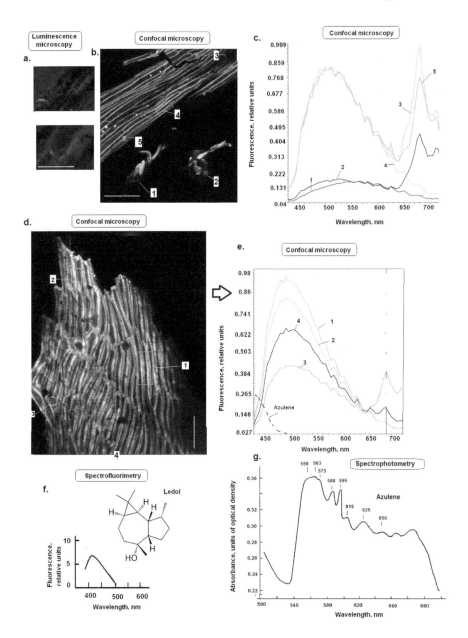

FIGURE 3.15 The fluorescence images and spectra of stem parts from Labrador tea, wild rosemary, *Ledum palustre* L. (a) Secretory cells (in secretory cavities) rich in the terpenoid alcohol ledol under Leica DM 6000B luminescence microscope on excitation by light at 340–380 nm. Bar = 200 μm. (b)–(e) Images of ROI and fluorescence spectra of the secretory cells recorded by Leica TCS SP-5 laser-scanning confocal microscope. Laser 405 nm. Bar = 200 μm; (f) fluorescence spectrum of ledol (2 mg/mL ethanol in 1 cm cuvette); (g) absorbance spectrum of azulene isolated from the object, purified by thin-layer chromatography, and dissolved in ethanol; its fluorescence spectrum is shown as a dashed line in (e).

L. palustre was observed, and the pigment azulene was found (Lynn et al. 1926). The fluorescence spectrum of the compound at 400–420 nm is represented as a dashed line in Figure 3.15e). Similarly, the pigment was specific to milfoil and wild ginger oils. Also, the phenol component may emit in blue, as shown in the emission spectra of intact tissue. Flavonoids such as quercetin and kaempferol, aromatic hydroxycinnamic acid, and chlorogenic acid have been identified as well as the presence of gallic and ellagic acids, tannin, and ellagitannins (Upyr et al 2015; Zhang et al. 2017).

3.13   *LEONURUS CARDIACA* L, COMMON MOTHERWORT

The genus *Leonurus* (family Lamiaceae), known as Leonuri cardiacae herba in Europe and Asia, includes species valuable for medicine: motherwort (*Leonurus cardiaca* L.) and *L. quinquelobatus* Gilib (syn. L. vilosus Desf.) (Wojtyniak et al. 2013; Shikov et al. 2014; Zhang et al. 2018). The plants, mainly common motherwort *L. cardiaca*, are used for their sedative effects in cardiovascular neurosis and hypertonia as well as for gastrointestinal symptoms and common stress states. The species have antibacterial, anti-inflammatory, and antioxidant activity and demonstrate a reduction of intracellular reactive oxygen species. The main biologically active components of the native drugs are flavonoid glycosides (rutin as the major component) and the alkaloid stachydrine (Murav'eva et al. 2007).

Zhang with coworkers (2018) summarized the structures of 259 compounds isolated from the genus *Leonurus*, featuring 147 labdane diterpenoids. Anti-inflammation is the major bioactivity discovered so far for the labdane diterpenoids from the genus *Leonurus*, whose further therapeutic potential still needs study. In addition to the traditional uterine contraction and sedative activity, recently, the cardiovascular protection effect of leonurine has attracted much attention. In addition, neuroprotection, anti-inflammation, anticancer, anti-platelet aggregation, and many other activities have been assigned to various compounds from the genus *Leonurus*. Among 70 bioactivity references cited, 57% of them concentrated on two alkaloids (leonurine and stachydrine), whereas only 20% were about the 147 diterpenoids. The sedative activity of *L. cardiaca* is thought to be due to the iridoids components just as in other anti-anxiety herbs. Some formulae of the most active compounds are represented in Figure 3.16.

In our experiments on autofluorescence with leaves and flowers of *L. cardiaca*, we saw bright blue or blue-green emission of secretory multicellular hairs excited by light in the 340–380 nm or 355–430 nm region as well as greenish-yellow under excitation by light 450–490 nm (Figure 3.16, images a and b). This fluorescence is usually specific to flavonoids and alkaloids (Roshchina 2008). In the red region, only the bases of the hairs weakly emitted due to chlorophyll. Confocal microscopy similarly demonstrated the predominant fluorescence in blue-green, although base cells of the hair show yellow emission (image b). Microspectrofluorimetry (spectrum c) showed maxima for the leaf hairs at 470 nm (possibly flavonoids) and 680 nm (chlorophyll), while in the petal of the flower, split maxima 470 and 500 nm of the secretory hairs were registered. An additional smooth maximum in the last case was observed at 610 nm, which may be due to some alkaloids or flavonoids. A monograph (Roshchina 2008) shows the fluorescence spectra of crystals

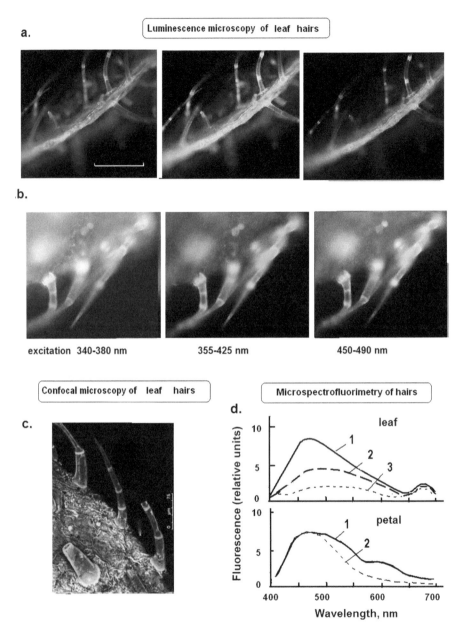

a.

Luminescence microscopy of leaf hairs

.b.

excitation 340-380 nm   355-425 nm   450-490 nm

Confocal microscopy of leaf hairs   Microspectrofluorimetry of hairs

c.   **d.**

FIGURE 3.16   Fluorescence images and spectra of *Leonurus cardiaca* L. (a) Luminescence microscopy (Leica DM 6000 B) images on leaf as a whole (upper side) and parts mainly with secretory hairs (lower side) after the excitation by light of different wavelengths. Bar=200 μm. (b) Confocal microscopy (Leica TCS SP-5) of leaf surface: summa of optical images after excitation by 488 nm laser. (c) Microspectrofluorimetry (MSF-1) of the secretory hairs after excitation by light at 360–380 nm. 1 (solid line): secretory hair, 2 and 3 (dashed lines): nonsecretory parts.

brightly emitting in red with the same maxima. Nonsecretory cells of the petal had no fluorescence. Biologically active alkaloids and flavonoids were identified by Zhang et al. (2018).

3.14   *MATRICARIA CHAMOMILLA,* WILD CHAMOMILE, GERMAN CHAMOMILE

Wild camomile or chamomile (*Matricaria chamomilla* L.), a well-known medicinal plant species from the Asteraceae family, is an example of predominant characteristic components in pharmaceutical material (Franke and Schilcher 2005; Murav'eva et al. 2007). Nowadays, it is a highly favored and much used medicinal plant in folk and traditional medicine (Duke 2002; Murav'eva et al. 2007; Singh et al. 2011) due to specific aroma components that are constituents of the essential oil from leaves and flowers. The plant species has multitherapeutic, cosmetic, and nutritional values. It is often used as a natural anti-inflammatory, antibacterial, and spasmolytic drug (Murav'eva et al. 2007; Singh et al. 2011). Moreover, the chamomile azulenes and proazulenes found in the essential oil have demonstrated antiallergic features (Singh et al. 2011). Some phenols and monoterpenes are included in the emission, but in a smaller degree in the glands and mainly concentrated in secretory hairs (Murav'eva et al. 2007; Singh et al. 2011).

Figure 3.17 represents our fluorescent analysis of fresh herbal material, flower, and leaf, from *M. chamomilla*. According to laser-scanning confocal microscopy, the white petal of the ligulate flower fluoresces in blue-green with maximum 480–500 nm, peculiar to flavonoids, from different ROI on the surface (a and b). An optical slice of anther with pollen grains demonstrated emission in blue-green (Figure 3.17c and d) with peaks at 520–530 nm on laser excitation at 405 nm. If a 488 nm laser was used, fluorescence was only in green (e). The observed emission may be due to flavonoids and flavins too. The spikes of pollen grains appeared to be brighter-emitting parts than other surfaces (c and e). As seen in fluorescence spectra (Figure 3.17d, spectra 1–5), the composition of compounds in the emitting petal and pollen varied from blue-green to yellow. This shows the contribution of flavins (at 520–530 nm, mainly, in the petal gland; spectrum number 1), carotenoids (530–540 nm, especially in pollen grains), coumarins (500–540 nm), and some flavonoids (460–480, 580–590 nm) (Roshchina 2008). Unlike the leaf, in the petals there was no chlorophyll in the glands, as seen from the spectra in (b). However, it should be noted that azulenes in the pollen may be bound to cell walls, and in this case, the pigments also contribute to red emission (maximum 620 nm) (Roshchina et al. 1995). According to fluorescence spectra recorded earlier by microspectrofluorimetry (Roshchina 2008), the pollen of chamomile may have maxima 435, 465, 570, and 620 nm, while the glands of the ligulate flower show maxima 460, 550, and 650 nm. The spectra differed from those of the leaf glandular hairs because of the lack of chlorophyll (maximum 680 nm).

Glandular hairs of the leaf that store valuable pharmaceuticals have been demonstrated to fluoresce mainly in the blue or green channels of confocal microscope (Figure 3.17f shows the sum of the optical slices, (g) on excitation by laser of 405 nm, while in the red channel of the apparatus, chlorophyll was seen (h). The highest emission was observed in the base part of the trichome. In the emission spectra from

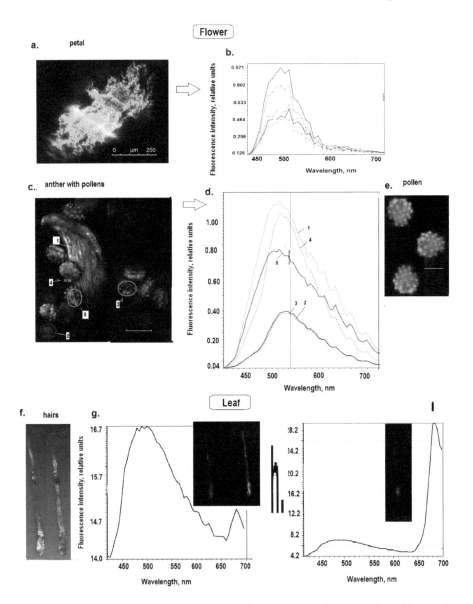

FIGURE 3.17   Fluorescence of terpene-containing secretory cells of hairs from *Matricaria chamomilla* L. Flower: (a) image of petal; (b) fluorescence spectra from various ROI on the surface recorded by Leica TCS SP-5 laser-scanning confocal microscope (laser excitation 405 nm); (c) image of anther with pollen and (d) their fluorescence spectra with the marked ROI (1–5); (e) image of pollen under 510 NLO Zeiss laser-scanning confocal microscope (laser excitation 488 nm). Leaf: (f) image of transverse optical slices of the fluorescing secreting hairs; (g) and (h) images of the hairs in three channels of Leica TCS SP-5 laser-scanning confocal microscope (laser excitation 405 nm) and their fluorescence spectra (marked by ROI in numbers), bar = 30 μm.

the parts of the hairs fluorescing in blue-green, a large maximum 500–510 nm was seen, while the small peak at 680 nm was due to chlorophyll (h). In the red channel, the green pigment emission predominated in the spectrum (h). In optical slices, via 1 µm (c), one can see the spreading of blue-fluorescing secretion along the multicellular hair filled with various fluorescing products. The emission may be due to azulenes or/and flavonoids predominating in *M. chamomilla* (Murav'eva et al. 2007; Singh et al. 2011). Isolated chamazulene (Roshchina and Roshchina 1993) from the species emitted at 425–430 nm (Roshchina et al. 2011), while most flavonoids emitted at 460–490 nm (Roshchina 2008). Glandular structures of the chamomile leaf fluoresced with maxima 460, 550, and 675 nm (Roshchina 2003, 2008), recorded by microspectrofluorimeter MSF 1. The use of microspectrofluorimetry gave more peaks in the fluorescence spectra than our measurement by confocal microscopy.

3.15 *MENTHA PIPERITA*, PEPPERMINT

In herbal medicine, perfumery, and cookery, mint, *Mentha piperita* from the Lamiaceae family, is widely used, mainly due to menthol giving a characteristic aroma to the essential oil (Murav'eva et al. 2007; Lawrence 2007; Kurilov et al. 2009; Elhoussine et al. 2011). The medicinal properties of tinctures are presented as natural drugs for spasmolysis in gastrointestinal diseases as well as in complexes of cardiovascular and antiseptic preparations. The herb is also added to tea. Thirty compounds were identified (Elhoussine derwich et al. 2011) in leaf oil, representing 58.61% of the total oil composition, and the major compounds in the leaves were terpenoids: menthone (29.01%), menthol (5.58%), menthyl acetate (3.34%), menthofuran (3.01%), 1,8-cineole (2.40%), isomenthone (2.12%), limonene (2.10%), α-pinene (1.56%), germacrene-D (1.50%), β-pinene (1.25%), sabinene (1.13%), and pulegone (1.12%).

Turner, Gershenzon, and Croteau (2000) actively studied the accumulation of essential oils in secretory structures of *Mentha piperita* by electron microscopy. The pattern of peltate glandular trichome initiation and ontogeny on expanding peppermint (*Mentha × piperita*) leaves has been defined by surveying the populations of peltate glands in each of seven developmental stages within sampling areas of leaf apical, mid-, and basal zones for both abaxial and adaxial surfaces. According to Murav'eva and coworkers (2007), various glands contain terpenoids.

In our experiments, the autofluorescence of glandular structures was analyzed. As shown in Figure 3.18 under transmitted light in a conventional microscope, the glands on the leaf appear brownish and rounded. At high magnification (objective ×40 or 60–100) and glycerin immersion, one can see that the glands consist of many different cells, which appear yellow-colored in the middle. When excited by UV or blue light, weak blue fluorescence is observed (images b, c, e), while the surrounding leaf surface and cells near the glands emit brightly. Under the confocal microscope (image g), in the two channels, including on the combined image (summa), there is a similar picture of the background. One can see maxima 470 and 530 nm as well as peak 680 nm due to chlorophyll in the fluorescence spectra measured from the gland by the microspectrofluorimeter (image f, solid line). Surrounded nonsecretory cells have only one maximum 680 nm (f, dashed line). The fluorescence spectrum of the

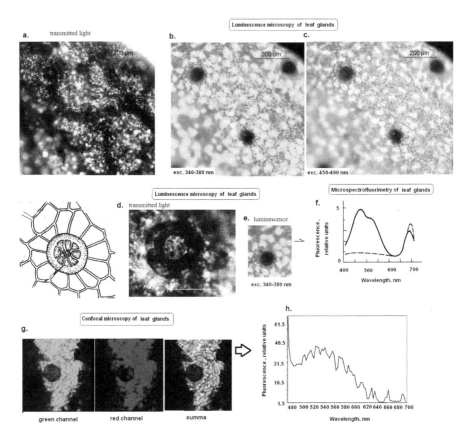

FIGURE 3.18   The images and fluorescence spectra of glands on peppermint, *Mentha piperita*, leaf. Seen in transmitted light of luminescence microscope Leica DM 6000 B (a, d) or at excitation by UV and violet light, microspectrofluorimeter MSF-1 (f; solid line, secretory cell; dashed line, non-secretory cell) or Leica TCS SP-5 laser-scanning confocal microscope, laser 488 nm, (g, h) bars = 100 µm.

gland was recorded using a confocal microscope (image g), and maximum 520–530 nm (spectrum h on image g) was shown after excitation by the 488 nm laser, based on the combined image of Figure 3.18d (summa).

3.16   *ORIGANUM VULGARE* L., COMMON ORIGANUM, WILD MARJORAM

Origanum vulgare L. (family Lamiaceae), the cultivated oregano culinary plant, is known as a medicinal and spice species (Bosabalidis 2002; Kintzios 2002b, 2003; Murav'eva et al. 2007; Azizi et al. 2009a, b; Shafiee-Hajiabad et al. 2014).

   This spice plant has antimicrobial, antifungal, insecticidal, and antioxidative effects on human health (Kokkini 1991; Kulisic et al. 2004; Bakkali et al. 2008). In folk and officinal medicine, it is included in pharmaceutical categories such as pectoral, carminative, diaphoretic, and so on (Murav'eva et al. 2007). As a medicinal plant, European oregano has traditionally been used as a carminative, diaphoretic, expectorant, emmenagogue, stimulant, stomachic, and tonic. In addition, it has been used as a folk remedy against colic, coughs, headaches, nervousness, toothaches, and irregular menstrual cycles (Kintzios 2002a). The active material is the essential oil, which consists of phenolic monoterpenoids (carvacrol and thymol up to 40%) and terpenoid (mono- and sesquiterpenes, where geranyl acetate is dominant at 5%) compounds (Murav'eva et al. 2007).

One of the noteworthy morphological characteristics of *Origanum* plants is the presence of glandular and nonglandular hairs (peltate trichomes or glandular scales) covering the aerial organs. Both types of hairs originate from epidermal cells (Netolitzky 1932). The glandular trichomes are numerous on the vegetative organs, such as stems, leaves, and bracts, while their density is reduced on the reproductive organs, such as calyces and corollas (Bosabalidis and Tsekos 1982). Bosabalidis and Kokkini (1997) have shown histological sections of the *Origanum vulgare* leaves, where the cavities are seen in capitate glandular structures. The glandular trichomes produce and secrete an essential oil with a characteristic odor, mainly due to the monoterpenes that are the major components of the oil (Scheffer et al. 1986).

According to our experiments, Figure 3.19 shows different emission of the glandular capitate trichomes at various wavelengths of excitation under a luminescence microscope (images a and b). In the fresh flower, the structures, being determined on transmitted light, look weakly fluorescent, if at all (except in image a, where emission was excited by light at 450–490 nm) against the background of surface emission in blue when excited in the UV. However, on blue (>455 nm) excitation of the surface, the cells of the glandular head rosette emit in yellow (image a). When the glandular heads have been shaken up from dried material (image b), more intense blue fluorescence in UV light is seen. When blue excitation (>450 nm) takes place, the whole head fluoresces in yellow. Under luminescence microscopy, the dried surface surrounding the dark sites of glandular heads also emits more intensely (image b) in comparison with fresh raw materials. In leaves, the secretory structures appear as different dark spots and spots with fluorescent rosette cells around dark spots. Yellow fluorescence of the rosette cells of glandular heads characterizes the presence of flavonoids—flavones such as kaempferol and quercetin or/and anthocyanins.

Using confocal microscopy, fluorescent pollen and pistil (c) and relative fluorescence spectra (d) may be observed. In all cases, pollen (spectrum 3) has maximum 480–500 nm, while the pistil itself (spectra 1 and 2) emits in both blue (peak 470 nm) and red (maxima 620, 650, and 680 nm) regions. Maxima in blue may be due to some flavonoids and aromatic acids (ROI 3 and 4) and those in red to anthocyanins and chlorophyll (ROI 1 and 2).

3.17   *SALVIA OFFICINALIS*, SAGE, GARDEN SAGE, COMMON SAGE, OR CULINARY SAGE

Sage, *Salvia officinalis* L., a perennial, evergreen subshrub, belongs to the mint family Lamiaceae and is known as an astringent, antibacterial, and anti-inflammatory agent in officinal and folk medicine (Kintzios 2001; Golovkin et al. 2001;

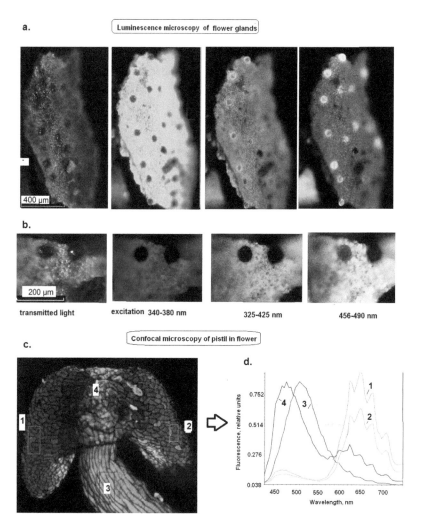

FIGURE 3.19 The images and fluorescence spectra of secretory cells of flower and leaf from *Origanum vulgare* L. emitted under various excitations of Leica DM 6000 B luminescence microscope (a and b) or Leica TCS SP-5 laser-scanning confocal microscope (c and d), laser 405 nm. (a) Part of fresh petal with sessile capitate glandular hairs clearly visible against the background of surrounding brightly emitting nonsecretory cells of the flower; (b) heads of individual glandular trichomes seen on dried material; (c and d) images and spectra of fresh pistil with pollen (ROI shown on images and respective spectra).

Duke 2002; Murav'eva et al. 2007; Hamidpour et al. 2014). It is often used for catarrh of respiratory ways (the airways), in particular for gargling of the pharyngeal cavity and stomatitis. This medicinal species contains the most active medicinal components against diabetes and inflammation (Duke 2002; Murav'eva et al. 2007; Jantová et al. 2014; Garcia et al. 2016). There is information about the antioxidant and antitumor properties of the plant (Garcia et al. 2016) and apoptosis in leukemia of mammalian cells (Jantová et al. 2014). The medicinal effects of the extracts are due to the essential oil of the herb, which contains the monoterpenes cineole, borneol, and thujone as well as tannins, flavonoids, and caffeic and chlorogenic acids. Essential oils may demonstrate anti-diabetic and anti-inflammatory features.

The autofluorescence of pharmaceutical materials changes during plant development, as is seen from images and fluorescence spectra (Figure 3.20). As shown in our experiments, the common fluorescence of the sage flower petal and its secretory hairs are clearly visible under a luminescence microscope (Figure 3.20a). For express analysis, changing the wavelength of the excitation, pharmacologists may observe the flavonoid blue emission in the UV (340–380 nm), which is quenched in the hairs under violet light (355–425 nm). The petal of *S. officinalis* contains many blue-colored anthocyanins, which also emit in the blue spectral region. The green

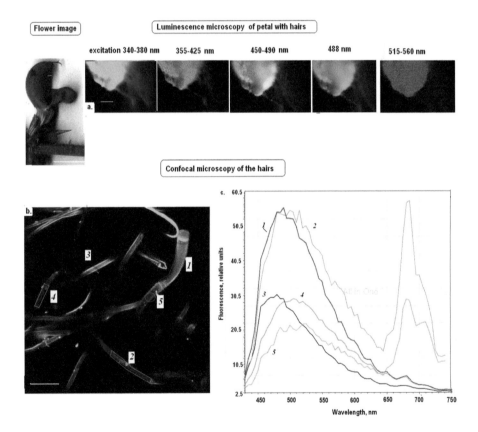

FIGURE 3.20 Fluorescence of phenol- and terpenoid-rich secretory cells of flower from *Salvia officinalis* L. (a) Part of petal with hairs emitting under various excitations of Leica DM 6000 B luminescence microscope, bar=30 μm; (b) and (c) numbered hair cells of petals fluorescing under Leica TCS SP-5 laser-scanning confocal microscope (laser excitation 405 nm) in green-yellow (bar=30 μm). In spectra 1 and 2, besides peaks 480–500 nm, maximum 680 nm is present. Other hair number 3 fluoresce.

fluorescence (510–520 nm) excited by light at 488 nm appears to be connected with flavins, and the yellow emission on excitation in the 450–490 nm region with carotenoids and some anthocyanins (Roshchina 2008). Finally, the bright red fluorescence on excitation in the 515–560 nm range is due to chlorophyll, which is usually masked by anthocyanins (excited in short-wavelength range) and is seen only under a luminescence microscope.

Using confocal microscopy, we studied the fluorescence and the emission spectra of petal secretory cells rich in various drugs (Figure 3.20b, c). There were different maxima in their fluorescence spectra (Figure 3.20c), which demonstrates the occurrence of different substances in the cells studied. In the spectrum of ROI number 3, the emission was in blue, which may relate to terpenoids fluorescing with maxima at 420–430 nm (Roshchina 2008). Samples number 1 and 2 had two maxima—one in blue-green at 490–500 nm and the second in red at 680 nm—showing the presence of flavonoids and chlorophyll. Spectra 3 and 4 display a lack of chlorophyll, unlike spectrum 5.

3.18 *TANACETUM VULGARE,* COMMON TANSY

Common tansy, *Tanacetum vulgare,* belonging to the family Asteraceae, is a flowering plant that used in medicinal, veterinary, and culinary uses. The button-like, roundish, flat-topped, yellow flower heads with a characteristic smell, like camphor with hints of rosemary, serve as a raw material for pharmacy (Rohloff et al. 2004; Murav'eva et al. 2007). Physicians recommend a tincture or decoction of the herb against *Ascaris* and pinworms for helminth invasion. Moreover, the species is included in various categories of treatment for problems with bile, in particular cholecystitis. In agriculture and veterinary medicine, tansy is used as an insecticide-bearing plant and cultivated as an insect repellent. The active components of the volatile oil include terpenes such as1,8-cineole, trans-thujone, camphor, and myrtenol, with the quantities and proportions of each varying seasonally. Among terpenoids, the species contains flavonoids and bitter compounds. Standardization of the raw pharmaceutical materials is carried out according to flavonoid content (no lower than 2.5% in terms of luteoline). *Tanacetum vulgare* demonstrates anti-inflammatory, antioxidant, antibiotic, and cytotoxic activity due to its essential oil (Coté et al. 2017). Tansy oil is a potent poison due to the presence of a high concentration of thujone, and even small doses can be fatal. It can also trigger hallucinations and severe nervous or neurotic disturbances while having addictive, narcotic effects. A total of 83 compounds were identified in both the oils that make up 98.4% of the oil in Estonia (Raal et al. 2014) The quantitatively most important components in tansy oil from Harju district were trans-chrysanthenone (41.4%), 1,8-cineole (9.6%), β-pinene (6.5%), α-pinene (5.0%), and 6-camphenone (4.6%). In the oil from Tartu district, β-thujone (47.2%) and trans-chrysanthenyl acetate (30.7%) predominated. Thus, the analysis resulted in assigning the *Tanacetum vulgare* of Harju and Tartu districts as trans-chrysanthenone and β-thujone-trans-chrysanthenyl acetate chemotypes. Tansy should not be used without skilled medical supervision. However, common tansy is rich in volatile, aromatic oils, and fresh young leaves and flowers may be used sparingly as a substitute in cooking. There is high variability in the constituents of the species (Piras et al. 2014; Muresan et al. 2015). The main volatile oil is thujone, a potent and bitter chemical often used medicinally as a wash to treat roundworm or internally to expel worms and cause abortions. Antimicrobial properties of tansy have also been demonstrated (Devrnja et al. 2017).

In experiments with the autofluorescence of the species, we noted that the tubular flowers of the plant have many glandular structures (Figure 3.21). The whole surface fluoresced in blue-green when excited by UV or violet light in a luminescence microscope, but the pistil was especially bright. The oval shape of glands is clearly seen on the surface as dark, weakly emitting spots. When the surface is excited by blue light at 455–490 nm, the secretory cell fluoresces in yellow, which enables the biserial structures of every gland to be visualized (image a). In the fluorescence spectrum of floral glands recorded by microspectrofluorimetry, there was maximum 490 nm and shoulder 550 nm after excitation by light at 360–380 nm, and pollen, as a secretory single cell, has peaks 490 and 600 nm. (Table 3.1). Confocal microscopy of floral glands (Figure 3.21b) shows in the spectra with a main maximum at 510–520 nm, analogous to green emission of leaf glandular hairs (Roshchina 2008).

3.19 SUMMARY OF FLUORESCENT SPECTRAL CHARACTERISTICS OF KNOWN MEDICINAL SPECIES

About 44 medicinal species have been studied by luminescence microscopy and microspectrofluorimetry and 20 species by confocal microscopy (Roshchina et al. 1997a, b, 1998a, 2004, 2007a; Roshchina 2003, 2004, 2005a, 2008, 2009, 2014). Some of the species are pharmaceutically valuable, and maxima in the fluorescence spectra of similar medicinal plants are shown in Table 3.1.

The data in Table 3.1 will give the reader a preliminary orientation regarding the chemical composition of secreting cells in fresh materials.

CONCLUSION

This is the first attempt to create a guide to pharmacy that can be used in laboratory practice and the education of students. The application of fluorescent methods to the express diagnostics of active fluorescent material is useful for the preliminary estimation of pharmacologically valuable materials. The knowledge of the drug composition of plant samples based on the analysis of native fluorescence or autofluorescence may help to select more potential species and their varieties for pharmacy. The atlas presented is devoted only to widespread medicinal plants, mainly specific to temperate climates. We hope that the list will increase due to the efforts of many scientists from all countries, especially from both tropical and arctic zones.

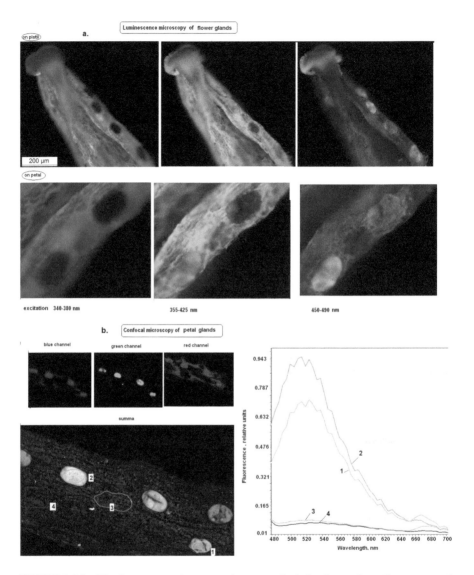

FIGURE 3.21   The fluorescence images and spectra of tubular flower from *Tanacetum vulgare*. (a) Luminescence microscopy (Leica DM 6000 B): weakly fluorescing glands among bright surrounding nonsecretory cells of pistil and tube surface bar = 100 μm. (b) Laser-scanner confocal microscopy (Leica TCS SP-5): views in three channels and sum of the fluorescence images, laser 458 nm. ROI on the summed image are related to the fluorescence spectra.

TABLE 3.1
Fluorescence Maxima of Fresh Secreting Plant Cells in Various Medicinal Plants

Plant Species (Latin Names)	Plant Species (English Names)	Secreting Cells and Secretions	Fluorescence Maxima (nm)
		Horsetails	
Equisetaceae			
Equisetum arvense L.	Field horsetail	Entrance of hydathode in tallus (epithem)	530, shoulder 475
		Drop of secretion from hydathode	460–470, 680
		Vegetative microspore	500
		Antheridium	465, 540, 680
		Spermatozoid	460, 540, 680
		Archegonium	500, 560
		Egg cell	520, shoulder 500, 585
		Ferns	
Polipodiaceae			
Dryopteris filix-mas (L) Scott	Shield fern	Soruces	475, 530, 680
		Seed-bearing plants	
Asteraceae			
Achillea millefolium L.	Common yarrow	Pollen	480
		Floral glands	465, 475, 500, 550
		Secretory cell of petal	565
		Leaf gland or secretory hair	450, 680
		Secretory cell of leaf	465, 545
		Secretory cell of root	460
Arctium tomentosum Mill (Schrank)	Cotton burdock	Secretory cell of leaf upper side	440 or 450, (500)550
		Secretion (crystal on the leaf surface)	440
		Secretory cell of leaf lower side	440
		Secretion (crystal on the leaf surface)	460
		Secretory hair of flower	465
		Nectary in red spots of petals	440
		Anther	465, 530
		Pollen	465–470
		Pistil	465, 550
Artemisia vulgaris L.	Mugwort, wormwood	Pollen	500
		Ligulate floral glands	465, 500, 550
		Leaf glands	450, 680
		Oil	460

(Continued)

TABLE 3.1 (CONTINUED)
Fluorescence Maxima of Fresh Secreting Plant Cells in Various Medicinal Plants

Plant Species (Latin Names)	Plant Species (English Names)	Secreting Cells and Secretions	Fluorescence Maxima (nm)
Calendula officinalis L.	Pot marigold	Pollen	460, 520, 580
		Secretory non-capitate hair of tubular flower	430, shoulder 500
		Secretory capitate hair of ligulate flower	475, 500
		Secretory cell of petal from ligulate flower	445, shoulder 530
		Secretory non-capitate hair of ligulate flower	450, 550
		Secretory capitate hair of sepal	445, 550
		Secretory non-capitate hair of sepal	440 or 440, 550, 680
		Secretory of calix sepal	435–440
		Secretory capitate hair of leaf	430
		Secretory non-capitate hair of leaf	560 or 585, 520, 600, or 500, 605
Echinops globifer (spharrocepha-lus) L.	Common globe thistle	Secretory cells of petal from red tubular flower	450, 680
		Secretory cells of petal from green tubular flower	460,680
		Pollen of red tubular flower	450
		Pollen of green tubular flower	470, 680
		Secretory cells of leaf	465, 550, 680
		Secretory hair of leaf	475, 550, 680
Matricaria chamomilla L.	German camomile, wild camomile	Pollen	435, 465, 570, 620
		Glands of ligulate flower	460, 550, 650
		Gland of leaf	460, 550, 675 or 450, 680
		Oil	440, 460
Tanacetum vulgare L.	Common tansy	Pollen	490, 600
		Floral gland	490, shoulder 550
Taraxacum officinale Wigg.	Common dandelion	Pollen	520
		Laticifer	550, 680
		Latex	465
		Secretory hair of leaf	465
Tussilago farfara L.	Common coltsfoot	Pollen	460, 520, 675

(Continued)

TABLE 3.1 (CONTINUED)
Fluorescence Maxima of Fresh Secreting Plant Cells in Various Medicinal Plants

Plant Species (Latin Names)	Plant Species (English Names)	Secreting Cells and Secretions	Fluorescence Maxima (nm)
Berberidaceae			
Berberis	European	Fruit peel	475, 520, 680
vulgaris L.	barberry	Pollen	465 (510), 540, 640
		Stigma of pistil	505
		Leaf secretory cell	465
		Flower secretory cell	535
Betulaceae			
Alnus glutinosa (L.) Gaertn.	Alder	Pollen	520
Betula verrucosa Ehrh.	European white birch, weeping birch	Pollen	440
		Exudate of bud scale	460–470
		Secretory hair of bud scale	500, 520
		Secretory cell of bud scale	480–495, 520, 680
Brassicaceae (Cruciferae)			
Capsella	Shepherd's purse	Secretory cell of leaf	500–520, 680
bursa-pastoris L.		Secretory hair of leaf	500–520, 680
Crassulaceae			
Bryophyllum	Bryophyllum	Pollen	475
daigremontianum Hamet et Perr		Floral nectar	460
Sedum hybrida L.	Stonecrop	Secretory cell of leaf	465 or 465, shoulder 550
		Secretory cell of root	475, 550
Cucurbitaceae			
Cucumis sativus L.	Cucumber	Pollen	475
		Stigma of pistil	475, 525, 680
		Secretory cell of flower petal	550–565, 680
		Secretory hair of the flower petal	475
Cucurbita pepo L	Gourd pumpkin	Pollen	530–540
		Stigma of pistil	450, 550, 680
Fabaceae			
Melilotus officinalis or *officinale* (L.) Pall	Yellow sweet clover	Pollen	480, 500–510
Trifolium pratense L.	Red clover	Pollen	500, 675
Trifolium repens L.	White clover	Pollen	515, 675

(Continued)

TABLE 3.1 (CONTINUED)

Fluorescence Maxima of Fresh Secreting Plant Cells in Various Medicinal Plants

Plant Species (Latin Names)	Plant Species (English Names)	Secreting Cells and Secretions	Fluorescence Maxima (nm)
Guttiferae			
Hypericum perforatum L.	Common St. John's wort	Pollen	460, 550
		Floral glands on petal	460, 680
		Leaf glands	440, 460, 680
		Exudate of leaf gland	440
Hippocastanaceae			
Aesculus hippocastanum L.	Common horse chestnut	Pollen	470, 680
		Secretory cell of bud scale	480, 680
		Secretory hair of bud scale	500
Iridaceae			
Colchicum autumnale L.	Common autumn crocus, meadow saffron	Pollen	470, 550
Lamiaceae (Labiatae)			
Lamium album L.	White dead nettle	Secretory cells of flower petal	475, 560–570
		Secretory cells of leaf	
Leonurus cardiaca L.	Motherwort	Secretory cells of leaf lower side	465, 680
		Secretory cells of leaf upper side	475, 680
		Secretory hair of flower	465, 600 or 465
		Secretion (crystal) in the hair	530, 600
Mentha piperita L.	Peppermint	Secretory cell of leaf	460, 530–540, 670 or 450, 525, 605
		Secretory hair of leaf	450, 595, 680
Salvia officinalis L.	Sage	Floral glands	465,550
		Oil	410
Malvaceae			
Hibiscus rosa-sinensis L.	Chinese hibiscus	Secretory hair on pistil	440
		Anther	475, 560
		Pollen	475, 560
Papaveraceae			
Chelidonium majus L.	Greater celandine	Pollen	540
		Latex	560
		Laticifer	480, 560, 680
Passifloraceae			
Passiflora coerulea L.	Blue-crown passion flower	Pollen	475, 570
		Stigma of pistil	475
		Extrafloral nectar	475, 550
		Extrafloral nectary	465–475, shoulder 515
		Floral nectary	475, 550
		Floral nectar	480, 545

(Continued)

TABLE 3.1 (CONTINUED)
Fluorescence Maxima of Fresh Secreting Plant Cells in Various Medicinal Plants

Plant Species (Latin Names)	Plant Species (English Names)	Secreting Cells and Secretions	Fluorescence Maxima (nm)
Pinaceae			
Abies sibirica L. (Ldb.)	Siberian fir	Resin duct of needle	465, 600
		Resin of needle	550
		Resinous cell of cone	465, 500, 600
		Secretory hair of cone	500
		Resin of cone	580
Larix decidua L.	Common larch, European larch	Pollen	460(675)
Larix sibirica Ledeb.	Siberian larch	Resin duct of needle	525
		Resin of needle	465
Picea abies Karst.	Norway spruce, common spruce	Resin duct of needle	525, 600
		Resin of needle	525, 600
		Resinous cell of cone	475, 575
		Resinous hair of cone	475, 525, shoulder 600
		Resin of cone	575
Picea excelsa (L.) Karst.	Spruce fir, Norway spruce	Resin duct of needles	430
		Resin of needle	550
		Resinous cell of cone	460
		Resinous cell in scale of cone	445–450
		Resin of cone	470
		Secretory cell of 3–10-day-old seedling	500, 580
		Cell in root meristem of seedling	485, shoulder 585
		Resin of seedling	450, 550
Picea pungens Engelm.	Colorado spruce, blue spruce	Resin duct of needles	440, shoulder 610
		Resin	550
Pinus sibirica Du Tour	Cedar pine	Resin duct of needle	465, 475
		Resin of needle	465
		Resinous cell of cone	475
		Resin of cone	500, 550
Pinus sylvestris L.	Scots pine	Resin duct of needle	450, 525–566
		Resin of needle	475–490 or 530
		Resinous cell of cone	485–490, shoulder 570
		Resinous cell of needle	475, 550
		Resin of cone	530
Plantaginaceae			
Plantago major L.	Ripple-seed plantain, great plantain	Pollen	540

(*Continued*)

TABLE 3.1 (CONTINUED)
Fluorescence Maxima of Fresh Secreting Plant Cells in Various Medicinal Plants

Plant Species (Latin Names)	Plant Species (English Names)	Secreting Cells and Secretions	Fluorescence Maxima (nm)
Rosaceae			
Crataegus oxyacantha L.	English hawthorn, European hawthorn	Pollen	480
Filipendula ulmaria (L.) Maxim	European meadowsweet	Pollen	480, 550
Rubus idaeus L.	Red raspberry, European raspberry	Pollen	430
Rutaceae			
Citrus sinensis (L.) Osbeck	Sweet orange	Secretory cell in peel of fruit	450, 560
Ruta graveolens L.	Rue	Pollen	Shoulder 475, 550
		Secretory glandular cell of leaf upper side	565
		Secretory cell of root (idioblast)	585–590
		Secretory hair of root	480, 575
		Secretory cell of root tip	500, 590
Salicaceae			
Populus balsamifera L.	Southern poplar	Pollen	470, 515
		Exudate of bud scale	470
		Secretory cell of bud scale	500, 680
		Secretory hair of bud scale	500
Salix virgata L.	Basket willow	Pollen	470, 515
Scrophulariaceae			
Symphytum officinale L.	Common comfrey, woundwort	Idioblast of leaf	535–556, 680
		Secretory hair of leaf	510
		Slime cell of leaf	535–540
		Slime	535
		Secretory cell of petal from flower	530, shoulder 465
		Stigma of pistil	530
		Water extract from leaf	440, 460
Solanaceae			
Capsicum annuum L.	Red pepper, cayenne pepper	Secretory cell of fruit peel	465, 530
		Secretory cell of fruit pericarp	525, 550
		Secretory cell of the seed stalk within fruit	475, 550

(Continued)

TABLE 3.1 (CONTINUED)

Fluorescence Maxima of Fresh Secreting Plant Cells in Various Medicinal Plants

Plant Species (Latin Names)	Plant Species (English Names)	Secreting Cells and Secretions	Fluorescence Maxima (nm)
Tiliaceae			
Tilia cordata Mill.	Little leaf linden	Pollen	470
Urticaceae			
Urtica dioica L.	Big stinging nettle, stinging nettle	Pollen	445 (680)
		Stinging hair	440, 460–470, 538, 680
		Water leachate	440
Urtica urens L.	Small nettle, dog nettle	Stinging hair	440, 470, 530, 680

Sources: Roshchina et al. 1997a, b, 1998a; Roshchina, V.V., *Fluorescing World of Plant Secreting Cells*, Science Publishers, Enfield, NH, 2008. With permission; Roshchina, V.V., unpublished data.

4 Fluorescence in Histochemical Reactions

Histochemical staining of plant raw materials may be useful as a method for studying the presence and localization of some drug compounds within cells and tissues or for the estimation of the viability of cells and the freshness of phytomaterials. Besides, some natural compounds formed under stress (xenobiotic or parasitic) may fluoresce and decrease the plant's pharmaceutical value. It should be noted that fluorescent dyes and fluorescent histochemical techniques for plants have not yet been much demonstrated. This chapter will cover the applications of some fluorescent dyes for studying stress and also compounds under consideration for pharmacy. To our regret, there is little information devoted specifically to the fluorescent histochemistry of major pharmaceuticals. A new guide book and review are devoted mainly to color reactions (Upton et al. 2011; Badria and Aboelmaaty 2019) but not to fluorescence. In the following, we shall consider histochemical methods for valuable substances such as neurotransmitters, phenols, lipids, and nucleic acids.

4.1 NEUROTRANSMITTERS

Acetylcholine, dopamine, norepinephrine (noradrenaline), epinephrine (adrenaline), serotonin, histamine, melatonin, and γ-aminobutyric acid, known as neurotransmitters, have been found not only in animals but also in plants (Emmelin and Feldberg 1947; Roshchina 1991, 2001; Murch et al. 2001; Murch and Saxena 2002; Murch 2006; Kulma and Szopa 2007; Ramakrishna et al. 2011a, b; Roshchina 2016a, 2019b; Erland and Saxena 2017, 2019; Arnao and Hernandez-Ruiz 2019). These molecules occur in all non-neural multicellular organisms as well as in unicellular organisms such as bacteria and protozoans (Roshchina 2010, 2016a). They should be termed *biomediators* (Roshchina 1989), involved in regulatory functions and in cell communication during the prenervous period. An evolutionary look at the role of neurotransmitters at all steps of evolutionary development—from unicellular organisms to multicellular ones (fungi, plants, and animals)—shows that non-synaptic functions of the substances arose earlier than synaptic functions in animals with a developed nervous system. So, the compounds may play a universal role as elementary molecular agents of irritation in any living cell (Roshchina 2010, 2016a). Today, neurotransmitters are considered as universal agents of communication that are instrumental in existing microorganism–microorganism, microorganism–animal, and microorganism–plant relationships.

Moreover, we have increasing evidence that neurotransmitters are multifunctional substances. Their concentrations in some organisms may be similar to those in animals and even higher in medicinal and food plant species (Konovalov 2019; Bhattacharjee and Chakraborty 2019). For common views of their functions, see Table 4.1. In plants, neurotransmitters play the roles of chemosignals: triggers of

TABLE 4.1

Content of Neurotransmitters in Plants and Established Functions of Neurotransmitters in Plants and Animals

Neurotransmitter	Content in Plants, μg/g of Fresh Mass	Functions in Plants	Functions in Animals
Acetylcholine	0.02–180	Regulation of membrane permeability and other cellular reactions up to growth and development in many plant species	Regulation of cell proliferation, growth, and morphogenesis. The carriage of nerve impulses across the synaptic chink from one neuron to another of impulses across the "motor plate" from a neuron to a muscle cell, where it generates muscle contractions.
Dopamine	0.1–547	Regulation of many cellular processes from growth and development to defense reactions	Decreases peripheral vascular resistance, increases pulse pressure and mean arterial pressure. The positive chronotropic effect produces a small increase in heart rate as well. Important for forming memories. In embryos of Vertebrata and lower animals, may regulate development
Noradrenaline	1–4000	Regulation of many cellular processes from growth and development to defense reactions	Increases peripheral vascular resistance, pulse pressure and mean arterial pressure as well as stimulating thrombocyte aggregation
Adrenaline	0.1–6760	Regulation of many cellular processes from ion permeability, growth, and development to defense reactions	Induced vasodilation (mainly in skeletal muscle) and vasoconstriction (especially skin and viscera)
Serotonin	0.22–3833	Regulation of growth and development of many plant cells	Control of appetite, sleep, memory and learning, temperature regulation, mood, behavior (including sexual and hallucinogenic), vascular function, muscle contraction, endocrine regulation, and depression. In embryos of Vertebrata and lower animals, may regulate development

(Continued)

TABLE 4.1 (CONTINUED)
Content of Neurotransmitters in Plants and Established Functions of Neurotransmitters in Plants and Animals

Neurotransmitter	Content in Plants, μg/g of Fresh Mass	Functions in Plants	Functions in Animals
Histamine	0.0017–4000	Regulation of growth and development under stress	Involved in many allergic reactions and increases permeability of capillaries; decreases arterial pressure, but increases intracranial pressure, which causes headache; smooth musculature of lungs contracts, causing suffocation; causes the expansion of vessels and the reddening of the skin, the swelling of cloth. Stimulation of the secretion of gastric juice and saliva (digestive hormone)

Sources: Konovalov, D.A., in *Neurotransmitters in Plants: Perspectives and Applications*, CRC Press, Boca Raton, FL, 2019; Bhattacharjee, P. and Chakraborty, S., in *Neurotransmitters in Plants: Perspectives and Applications*, CRC Press, Boca Raton, FL, 2019; Roshchina, V.V., *Biomediators in Plants. Acetylcholine and Biogenic Amines*, Biological Center of USSR Academy of Sciences, Pushchino, 1991; Roshchina, V.V., *Neurotransmitters in Plant Life*, Science Publishers, Enfield, NH, 2001; Roshchina, V.V., *Model Systems to Study Excretory Function of Higher Plants*, Springer, Dordrecht, Netherlands, 2014; 2016.)

growth and development (seed and pollen germination, embryo formation, etc.); regulators of ion permeability, energetics, metabolism, and movement (dealing with roots, leaves, and stomata cells); and protectors against stresses as well as participating in intercellular and intracellular relations. There is important literature in which experimental data are combined (Hartman and Gupta 1988; Tretyn and Kendrick 1991; Roshchina 1991, 2001; Wessler et al. 2001; Kulma and Szopa 2007; Ramakrishna et al. 2011a; Erland and Saxena 2019; Ramakrishna and Roshchina 2019; Roshchina 2019b). The stimulatory effects of some neurotransmitters on plant growth reactions may be useful in agriculture (Bamel and Gupta 2019; Sharma and Gupta 2019; Erland and Saxena 2017). Moreover, the compounds within plants play important medicinal roles in regulating key human processes (Table 4.1) and are recommended for treatment, for instance in Alzheimer's disease (Bhattacharjee 2019; Bhattacharjee and Chakraborty 2019).

The interest of physicians and pharmacologists in the detection of neurotransmitters in plants is linked with the fact that they are contained in significant concentrations in food and medicinal plant raw materials. Today, there is the potential to use medicinal plants rich in some neuromediators for pharmacology and medicine as alternatives to chemically synthesized drugs.

4.1.1 ACETYLCHOLINE AND CHOLINESTERASE

Acetylcholine, formed from choline and acetic acid, plays an important role in neuronal impulses in animals and humans; therefore, its earlier identification in plant raw material is valuable for pharmacy. It has been identified in more than 80 plant species from 38 families (Fluck and Jaffe 1974; Tretyn and Kendrick 1991; Roshchina 1991, 2001; Wessler et al. 2001). The amounts of the neurotransmitter acetylcholine in some medicinal plant cells may achieve high values (Roshchina 1991, 2001). For example, these concentrations are especially marked in leaves and fruits of *Artocarpus champeden*, *Crataegus oxyacantha*, and *Urtica urens*. In stinging hairs of *Urtica dioica* and *Urtica urens*, the amount of neurotransmitter is as high as 10^{-1} M or 120–180 nmol/g fresh weight. Kawashima with coauthors (2007) found that at the top of the bamboo shoot, the acetylcholine content was approximately 80 times higher (2.9 mmol/g) than in the rat brain. The compound is found in extracts from plant tissues not only in free form, but also in various conjugates (Fluck et al. 2000). Acetylcholine (0.05–0.15 g per day) decreases arterial pressure, acts as a vasodilator, retards the cardiac rhythm, induces the stenosis of pupils, and enhances the contraction of smooth musculature in internal organs. Most medicinal effects are due to the presence of the neurotransmitter and the balance between its concentration and the activity of cholinesterase (the enzyme of its catabolism). Moreover, the amount of neurotransmitter shows how acetylcholine can regulate plant growth and morphogenesis, which is very important for cultivation in agriculture (Bamel and Gupta 2019). Mitigating effects of acetylcholine on seed germination were also observed in plants growing under osmotic stress (Braga et al. 2017).

The localization of acetylcholine in plant tissues and cells is usually determined based on the presence of a cholinesterase, the enzyme that catalyzes its hydrolysis. Cholinesterase/acetylcholinesterase activity serves as a marker of acetylcholine (Kakariari et al. 2000). The enzyme is found in different plant tissues depending on species. In corn, acetylcholinesterase is predominantly localized in vascular bundles (Yamamoto and Momonoki 2012). On the cellular level, cholinesterase has been found in the cell wall, exine of spores and plasmalemma (Fluck and Jaffe 1974) as well as in the nucleus (Maheshvary et al. 1982), chloroplasts (Roshchina 1989, 1991, 2001), and vacuoles (Roshchina 2018).

Today, the main indirect way to determine acetylcholine in plant cells using a fluorescent method is to assay the presence of cholinesterase (Figure 4.1). One may determine the reactions with coumarinylphenylmaleimide (Parvari et al. 1983) or with the red analog of Ellman reagent (2,2-dithio-bis-(p-phenyleneazo)-bis-(1-oxy-8 -chlorine-3,6)-disulfuric acid in the form of a sodium salt) for a fluorescent product with thiocholine (Roshchina et al. 1994; Roshchina 2001). The first candidate for histochemical application was the flavonoid derivative coumarinylphenylmaleimide (Parvari et al. 1983), which has a maximum of 390 nm in the excitation curve and 473 nm in the emission spectrum. After treatment with the second reagent, a blue compound was formed on the surface of pollen from various species and fluoresced with a maximum of 500–510 nm (Roshchina et al. 1994; Roshchina 2001), while

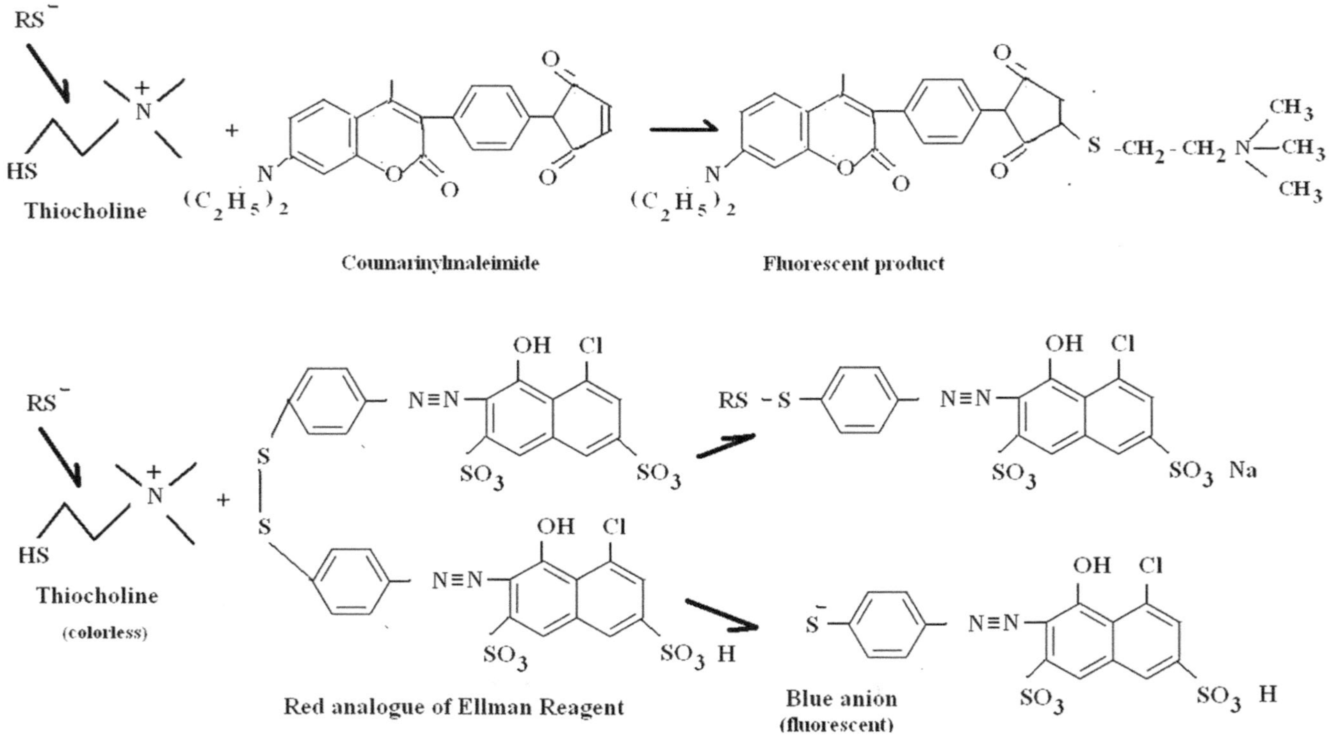

FIGURE 4.1 Possible schemes for formation of fluorescent products in histochemical reactions for cholinesterase activity. (From Parvari, R., et al., *Anal. Biochem.*, 133, 450–456, 1983; Roshchina, V.V., 2001. *Neurotransmitters in Plant Life*, Science Publishers, Enfield, NH, 2001.)

untreated cells of *Hippeastrum hybridum* pollen, for example, usually emitted with a small maximum of 480 nm if at all.

An indirect histochemical assay may be the reaction of cholinesterase *in situ* with special fluorescent inhibitors of cholinesterases, such as berberine or chelerythrine, with maximum emission at 540 nm (Roshchina 2007a, b, 2008). Both compounds bind to the cell surface, where the enzyme is usually localized.

An indirect fluorescent method for the determination of cholinesterase activity in the medicinal species *Artemisia tridentate* Nutt. (Turi et al. 2014) was applied using the artificial reagent 10-acetyl-3,7-dihydroxyphenoxazine (Amplex® Red reagent), a sensitive fluorogenic probe for hydrogen peroxide. Acetylcholinesterase converts the acetylcholine substrate to choline, which is then oxidized by choline oxidase (present in the mixture) to betaine and hydrogen peroxide. The emission is observed at 590 nm (excitation 530–560 nm). It should be noted that this fluorescent method has not yet been used in plant histochemistry.

Some medicinal plants are rich in acetylcholine and have active cholinesterase (Tretin and Kendrick 1991; Roshchina 1991, 2001; Konovalov 2019). These are *Alangium lamarckii* Thw., *Liquidambar styraciflua* L., *Amaranthus caudatus* L., *Rhus copallina* L, *Ipomoea abutiloides* (Carnea), *Equisetum robustum* A. Braun ex Engelm., *Cnidoscolus texanus* (Muell. Arg), *Codiaeum variegatum* (L.) Blume, *Urtica dioica* L., etc. Only the genus *Urtica* is in pharmaceutical nomenclature; the other species are known as food plants or are used in folk medicine and homeopathy. In their role as drugs, physicians (ophthalmologists) mainly recommend the use of anticholinesterase compounds in order to increase the concentration of acetylcholine in the eye. Signaling and restoration of the functioning of the neuromuscular system by the neurotransmitter have not been estimated yet for extracts from fresh plant pharmaceutical material rich in the compound. It should be noted that another direction of application in pharmacy may be a search for native species containing the anticholinesterase compounds (Murray et al. 2013). This is a real problem in the treatment of Alzheimer's disease; efforts are being made in search of new molecules with anticholinesterase activity. Naturally occurring compounds from plants are being considering as potential sources. New inhibitors have led to the discovery of an important number of secondary metabolites and plant extracts with the ability to inhibit the enzyme, increasing the levels of the neurotransmitter acetylcholine in the brain and thus improving cholinergic function in patients with Alzheimer's disease and alleviating the symptoms of this neurological disorder. Special attention is being paid to essential oils with anticholinesterase activity related to sesquiterpenes (Owokotomo et al. 2015). For example, essential oil extracts from leaf, seed, stem, and rhizome of four medicinal plants—*Aframomum melegueta* K. Schum, *Crassocephalum crepidioides* (Benth S. More), *Monodora myristica* (Gaertn.), and *Ocimum gratissimum* (Linn)—were tested for acetylcholinesterase inhibitory activity using Ellman's colorimetric method and compared with the known acetylcholinesterase inhibitor galantamine as a reference, and essential oils from cultivated medicinal plants such as *Salvia officinalis*, *Mentha piperita*, *Melissa officinalis*, and others, demonstrated high inhibitory activity (over 80%) in many probes (Ornan et al. 2008). The importance of extracts from acetylcholine-rich species is also a new field in medicine.

4.1.2 BIOGENIC AMINES

Biogenic amines are metabolites derived from common amino acids—phenylalanine, tryptamine, and histidine—as a result of the decarboxylation and hydroxylation of the corresponding amino acids. For plants, similar data have been summarized in some monographs and reviews (Roshchina 1991, 2001; Kulma and Szopa 2007; Ramakrishna et al. 2011; Erland et al. 2016; Erland and Saxena, 2017). The occurrence of biogenic amines such as catecholamines, serotonin, and histamine in plant cells (Werle and Raub 1948; Marquardt and Vogg 1952; Kutchan et al. 1988; Barwell 1979, 1989; Roshchina 1991, 2001, 2010; Kuklin and Conger 1995; Ekici and Coskun 2002; Kulma and Szopa 2007; Erland and Saxena 2017) and the sensitivity of every living cell (microbial, plant, and animal) to the neurotransmitters, especially in the aspect of the growth and development in biocenosis (Roshchina 2010), is of interest to the wider circle of biologists. The knowledge is valuable for fundamental biochemistry and ecology as well as for medicine, pharmacology, and food production. Neurotransmitters play a significant role in human health and medicine as drugs; for example, biogenic amines. In most cases, it would not be possible to see the location of the biogenic amines within individual cells and their secretions by using chemical methods that require large amounts of raw material and preliminary liberation from proteins, amino acids, and other amines from the cells and tissues. Biogenic amines play various important roles in the physiology and development of eukaryotic cells; they are sources of nitrogen and precursors for the synthesis of hormones, alkaloids, nucleic acids, and proteins (Ekici and Omer 2019).

In plants, the biogenic amines in raw vegetables are usually present both in free form and conjugated with other molecules, such as phenolic acid and proteins, with levels depending on variety, ripening stage, and storage conditions. Different plant foods and pharmaceutical preparations contain various biogenic amines that are pharmaceutical materials (Konovalov 2019; Ekici and Omer 2019). As shown in Table 4.2, pure individual amines are used in medicine.

The positive medicinal role of catecholamines in food for humans focuses on several physical and mental disorders: for example, Alzheimer's disease, Parkinson's disease, schizophrenia, glaucoma, Huntington's disease, epilepsy, arrhythmias, and so on. The possible application of phytopreparations rich in the compounds has potential for officinal and alternative medicine. Dopamine's functions are not limited to the central nervous system. Dopamine is a coregulator of the immune system (Atkinson et al. 2015), tissues, and organs (Zhang et al. 2012). Disturbances in the dopaminergic system cause many health problems, including high blood pressure, mental disorders (e.g., schizophrenia), and neurodegenerative diseases (e.g. Parkinson's disease). The hypothesis of Nalbandyan (1986) showed the possible ability of catecholamines to bind copper-containing proteins of the brain, in particular neurocuprein, known as a vasoactive compound. In this case, through redox reactions, it may induce neurodisorders such as schizophrenia.

Methods for the detection of biogenic amines in plants may be biological (based on the sensitivity of muscle contraction to plant extracts containing the compound studied) or chromatographic (to separate similar extracts and identify compounds

TABLE 4.2
Medicinal Uses of Individual Biogenic Amines

Dopamine	Noradrenaline	Adrenaline	Serotonin	Histamine
In shock states and acute cardiac insufficiency	In shock states and acute cardiac insufficiency	Stops bronchial asthma and other acute allergic reactions. Used in care of glaucoma and as vasoconstrictive and anti-inflammatory agents in otorhinolaryngologic and ophthalmic practices. Acts on blood vessels and as a hormone.	(5–10 mg per day) is useful against hemorrhagic syndromes and increases capillary stability and decreases hemorrhage in anemia.	Included in drugs against polyarthritis and rheumatism, and in small doses ($\sim 10^{-7}$ M) in order to prevent acute allergic reactions. Stimulates secretion of gastric glands and spastic contraction of enteric musculature. Histamine exerts its effects by binding to cell membrane receptors of skin and respiratory, cardiovascular, gastrointestinal, and immunologic reactions.

Sources: Ekici, K. and Omer, A.K., in *Neurotransmitters in Plants: Perspectives and Applications*, CRC Press, Boca Raton, FL, 2019; Bhattacharjee, P. and Chakraborty, S., in *Neurotransmitters in Plants: Perspectives and Applications*, CRC Press, Boca Raton, FL, 2019.

by mass spectrometry). For the study of their location, immunological methods (Ramakrishna et al. 2011a) and fluorescent methods (Kimura 1968; Kutchan et al. 1998) have been used.

Some comments are required regarding the significance of biogenic amines for plants themselves. Dopamine has been found so far in 40 species of 23 families (Roshchina 1991, 1992, 2001; Kulma and Szopa 2007). Its medicinal importance is connected with food and pharmaceutically valuable species. The greatest concentration of dopamine is contained in the peel of banana fruit; in the pulp, it is 87 times lower. Noradrenaline was identified first in *Banana* fruits and considered responsible for the therapeutic effect (Waalkes et al. 1958; Wagner 1988). In the leaves of some leguminous Fabaceae and Urticaceae, its amount is approximately 4000 nmol/g fresh mass. The amount of dopamine found varies during plant development (Kamo and Mahlberg 1984) and sharply increases during stress (Swiedrych et al. 2004). Dopamine is found in all organs of medicinal species *Aconitum napellus* (family Solanaceae), but especially in flowers and young fruits, whereas its concentration is lowest in roots (Faugeras et al. 1967). In latex of poppy *Papaver bracteatum*, the content of dopamine is 0.1% (1 mg/mL), and it easily transforms into morphine and other alkaloids (Kutchan et al. 1986). As well as dopamine itself, there are derivatives of dopamine in plants, for example, dopamine betaxanthin in *Portulaca oleracea* (Gandía-Herrero et al. 2005a). The important role of the catecholamines in plant fertilization as well as in fruit and seed development has been described (Roshchina 2001, 2014). The quantity of dopamine sharply increases in response to wounding and stress. After the wounding of cactus *Carneginea gigantea*, a defensive callus develops, and its cortical tissue (pulpy cortex) contains dopamine as a main phenolic component (Steelink et al. 1967). The concentration of dopamine after wounding in the pulp of the callus can be as high as 1%.

The accumulation of catecholamines in leaves of some plants may relate to stress. Furthermore, the noradrenaline level in the peel of banana fruits increased from 5.1–6.2 in unmatured and yellow matured to 15.2 µg/g of fresh mass in yellow-black supermatured fruit, and the increase in the pulp from 2.5 in yellow matured to 10.1 µg/g of fresh mass in supermatured fruit is associated with oxidative processes during ripening (Foy and Parratt 1960). Feng with coauthors (1961) found a high concentration of noradrenaline in the medicinal plant *Portulaca oleracea*. In the ripening or overripened state, the concentration of noradrenaline increases two- to fourfold compared with unripe fruits. In plants that are able to respond to mechanical stimuli with motor activity, such as *Albizia julibrissian* and *Mimosa pudica*, noradrenaline accumulates in motor organs and axes of leaves (Applewhite 1973). Dopamine has an exogenous defensive reaction to stress connected with alkali salt damage (Jiao et al. 2019) or drought (Gao et al. 2019).

Another important biogenic amine, serotonin (5-hydroxytryptamine), has been identified in more than 130 species of 48 families (Roshchina 1991, 2001; Ramakrishna et al. 2011a; Erland et al. 2016; Erland and Saxena 2017, 2019). The amount of serotonin varies depending on the taxonomic position of the plant (family, species, and cultivar) and the part of edible tissue analyzed. In addition, it has been found that the serotonin levels in plants are affected by the environment, including light levels, stresses (biotic and biotic), growth conditions, vegetative growth, reproductive development, seasonal rhythms, unfavorable location, and pathogens

(Erland and Saxena 2017). As well as free serotonin, its conjugated derivatives such as N-feruloylserotonin, N-(p-coumaroyl) serotonin, and N-(p-coumaroyl) serotonin mono-β-D-glucopyranoside were isolated from plants. Serotonin interacts with the phenylpropanoid pathway by serving as an amine substrate for the formation of hydroxycinnamic acid amides by enzyme-mediated condensation of a phenolic substrate (e.g., coumaric acid and ferulic acid) to an amine substrate (serotonin) (Macoy et al. 2015; Erland et al. 2016). Immunolocalization analysis performed by S. Kang and coauthors showed that a significant amount of serotonin is contained in the xylem cells of conductive bundles of aging leaves (Kang et al. 2007; Ramakrishna et al. 2011a, b). According to many reviews (Ekici and Omer 2019; Konovalov 2019; Bhattacharjee and Chakrabothy 2019), the fruits and seeds of many plants contain over 100 times more serotonin than vegetative organs, which is comparable to its concentration in animal tissues. The greatest amount of this substance in the banana is found in the peel. In the pulp, the amount of serotonin is considerably lower.

For food and medicinal plants in pharmacy, it is necessary to determine the presence of biogenic amines by suitable fast diagnostic methods, because the compounds may be dangerous to people's health, especially when their amounts reach a critical threshold (Ekici and Omer 2019). Since plants undergo continuous exposure to various biotic and abiotic stresses in their natural environment, the raw pharmaceutical material should be controlled for increased content of biogenic amines. Unfavorable growth conditions stimulate the accumulation of biogenic amines in increased amounts, which is dangerous for both human and plant health. High concentrations of the substances frequently induce intoxication and intolerance in humans, which affects cell proliferation and differentiation, the transmission of nerve impulses across a synapse, control of blood pressure, allergic responses, and cellular development control.

In the next subsection, we shall consider the use of fluorescence as applied to histochemical methods for the fast assay of biogenic amines in plants.

4.1.2.1 Fluorescence in Histochemical Analysis of Biogenic Amines

By the 1990s, the presence of dopamine had been established by electron microscopy in vacuoles and vesicles in *Papaver somniferum* latex with alkaloids (Roberts et al. 1983; Roberts 1986; Kutchan et al. 1986). As early as 1968, Kimura applied a fluorescent method for this purpose to plants based on the approach usually studied in animal physiology. This was visible emission from fluorescent products formed as a result of the reaction with certain neurotransmitters. Similar histochemical reactions characteristic for biogenic monoamines are shown in Figure 4.2. The basis of the mechanisms of fluorophore formation (Falck 1962) in histochemical reactions with formaldehyde, glyoxylic acid, or *o*-phthalic aldehyde is the Pictet–Spengler reaction (considered as a special case of the Mannich reaction), in which a β-arylethylamine undergoes ring closure after condensation with an aldehyde or ketone. After heating, even in physiological conditions, some fluorescing isoquinolines and carbolines are formed, which are used in histochemical assays. In freeze-dried tissues, for example, formaldehyde converts catecholamine and serotonin to 3,4-dihydroisoquinolines and 1,2,3,4-tetrahydro-β-carboline, which show green and yellow emission, respectively, in ultraviolet (UV) actinic light. It also reacts

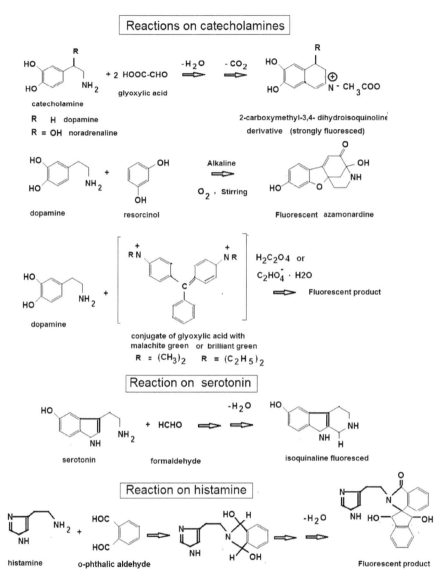

FIGURE 4.2 Main scheme of the formation of fluorescent products in histochemical reactions for biogenic amines. (Sources: Roshchina, V.V. *J. Fluoresc.*, 26 (3), 1029–1043, 2016; Zhang, X. et al. *Anal. Chim. Acta*, 944, 51–56, 2016; Roshchina, V.V. In: *Neurotransmitters in Plants: Perspectives and Applications*, Eds. A. Ramakrishna and V.V. Roshchina, pp.135–146, Boca Raton: CRC Press, 2019, and unpublished data.)

with catecholamines (dopamine, noradrenaline, and adrenaline) forming fluorescent products with the emission maximum at 460–475 nm. The reactions also occur after treatment with glyoxylic acid, and an emission maximum of the products with catecholamines is seen at 470–475 nm.

Zhang and coworkers (2016) have presented a new method to determine catecholamines *in situ* with the example of dopamine. There was a rapid reaction (5 min) between resorcinol and catecholamine, which resulted in the formation of fluorescent azamordine with high quantum yield. The method shows high sensitivity, with the limit of detection at 1.8 nM dopamine.

The determination of biogenic amines was earlier carried out with large samples of plant materials using treatment with trichloroacetic acid or hydrochloric acid without or with column chromatography (Kuruma et al. 1994; Ekici and Coskun 2002), but fluorescent histochemical methods need only small amounts of sample. The difficulty was overcome in experiments on animals, where special procedures of drying at 80 °C in histochemical assays with induced fluorescence were applied (Eninger et al. 1968). Kimura (1968) discovered the presence of serotonin in some vegetables and fruits using histochemical methods. Parenchymal epidermal cells of banana fruit and epidermal cells of onion showed the characteristic yellow fluorescence of serotonin. This was observed in the tissues of the leaves and stems of the plant. However, peel cells of lemon did not have similar properties (Kimura 1968).

The histochemical fluorescent methods included molecular probes: glyoxylic acid as a reagent for catecholamines such as dopamine, noradrenaline, and adrenaline (Axelsson et al. 1973; Lindvall et al. 1974; Markova et al. 1985) and *ortho*-phthalic aldehyde as a reagent for histamine (Cross et al. 1971; Douabale et al., 2003). Both molecular probes form fluorescent products—derivatives of isoquinolines. The fluorescence of intact animal cells treated with glyoxylic acid shows maxima at 480–500 nm (sometimes at shorter wavelengths) for common catecholamines. The presence of biogenic amines in plant cells and cellular secretions may be analyzed by histochemical staining with glyoxylic acid as a reagent for catecholamines and *o*-phthalic aldehyde as a reagent for histamine based on the methods presented in animal physiology. Both molecular probes form fluorescent products—derivatives of isoquinolines (Figure 4.2). The reaction with histamine leads to the formation of a fluorescent product complex including several derivatives of imidasolyl ethylamine. The fluorescence of the complex is inhibited in acidic medium and emits in alkaline medium, although it is unstable, but this can be overcome by drying. The fluorescence of intact animal cells treated with glyoxylic acid shows maxima at 480–500 nm for the total sum of catecholamines, while after the addition of *o*-phthalic aldehyde, the maximum is 430–450 nm, which is specific to histamine. At lower concentrations, 10^{-8}–10^{-7} M of histamine, the emission of animal tissue slices in the reaction under UV light in a luminescence microscope is blue, while at higher concentrations, the complex of fluorescing products emits in green and at maximum concentrations, in yellow. As well as the described methods, the determination of dopamine in pollen grains from *Dolichothele albescens* was carried out with formaldehyde vapor, based on analogy with the conjugate of glyoxylic acid with malachite green (Guidry et al. 1999). For the histochemical assay of catecholamines, we also tried to use the conjugate of the same acid at 1% with brilliant green (Figure 4.2), and red fluorescence of the cells was seen (Roshchina 2016b). Formaldehyde vapor (at heating up for drying) also induced an increase in fluorescence at 500–520 nm in pollen of *D. albescens*, although the method appears not to be specific only for catecholamines, because emission in red can be due to serotonin as well.

Unlike the above-mentioned assays with glyoxylic acid or o-phthalic aldehyde, for determination of the presence of serotonin, only formaldehyde is used. This aldehyde causes fluorescence in green-yellow and is used for the analysis of some food plants (Kimura 1968).

4.1.2.2   Visualization of Biogenic Amines

The preparation of samples includes the following procedures. Living cells on slides (subject glasses) are put into Petri dishes and treated with aqueous solutions of reagents. For the determination of catecholamines, 1% glyoxylic acid is applied according to the method for animal embryos (Markova et al. 1985) as shown by Cross et al. (1971) and Barwell (1989). To analyze the presence of histamine, the samples are treated with 0.5% o-phthalic aldehyde. After 10–20 minutes of staining with the reagents, they are dried at 50–80 °C. According to the fluorescent method of Kimura (1968), analogous to experiments on animals, green-yellow emission could be seen after the treatment of plant cells with formaldehyde vapor.

The fluorescence of living cells and isolated organelles was observed and photographed using a Leica DM 6000 B luminescence microscope (Roshchina et al. 2015a, b). The first work in this direction was done on intact pollens and isolated organelles. For approximate estimation of the amount of biogenic amines in individual cells, dried films of dopamine or histamine at different concentrations from 10^{-8} to 10^{-2} M on slides were prepared for the composition of concentration curves. The fluorescence of intact cells under a luminescence microscope was excited by UV (filters UVS-6 with maximal transmission at 360–370 nm and SZS-7 with maximal transmission 380 nm or violet light 405–436 nm) and photographed under a Leica DM 6000 B luminescence microscope, which allows visualization of the emission at an excitation of 340–380 nm (filter system A: dichromatic mirror 400 nm; suppression filter LP 425 nm).

The emission intensity (I) of intact cells excited by UV (360–380 nm/360–405 nm) or violet light (420–435 nm) of intact cells was recorded in two different spectral regions at 450–460 nm (I_{450}) and 520 nm (for samples with very high concentrations of catecholamines or histamine). Plant cells were analysed in subject glasses put in Petri dishes according to method for determination of catecholamines described for animal embryo cells (Markova et al. 1985). This was applied to plant samples such as pollen, vegetative microspores of spore-bearing species (Roshchina and Yashin 2014; Roshchina et al. 2015a), secretory cells of *Urtica dioica* (Roshchina and Yashin 2014; Roshchina 2014) and the alga *Chara vulgaris* (Roshchina et al. 2019). It is possible to determine histamine in plant cells by treatment with 1% or 0.5% o-phthalic aldehyde, according to Cross and coworkers (1971) for animals and Barwell (1989) for algae; this has been applied to plant generative cells (pollen and spores of *Equisetum arvense*) and secretory cells of *Urtica dioica* (Roshchina and Yashin 2014; Roshchina 2014; Roshchina 2019a, b, c).

The chloroplasts and nucleus appear as blue organelles on staining with glyoxylic acid and as blue or yellow ones under o-phthalic aldehyde (Roshchina et al. 2016a). Red fluorescence of plastids at 680 nm (specific to chlorophyll) decreased due to an increase in blue or blue-green emission. After the treatment with o-phthalic aldehyde, the cells fluoresced like animal cells. In germinating pollen, the fluorescent spermia nuclei moved along the tube. Thus, DNA-containing organelles such as nuclei and chloroplasts in vegetative microspores of *Equisetum arvense* or nuclei in pollen of *Hippeastrum hybridum* emitted as separate structures in the cells.

New data will be given in the following, especially for secretory cells of medicinal plants. In Figure 4.3, there are fluorescent images and fluorescence spectra of some secretory structures of medicinal plants treated with histochemical reagents for dopamine and other catecholamines. The reader can see the results of the histochemical analysis of secretory cells of known medicinal phytopreparations, such as buds of *Betula verrucosa* Ehrh, stinging hairs of *Urtica dioica* L. on leaves, leaves of *Eucalyptus cinerea* F.Muell., and root idioblasts of *Ruta graveolens* L. Blue and green-yellow emission are seen after staining with glyoxylic acid in Figure 4.3. The buds of *B. verrucosa* are used as natural diuretic and cholagogue drugs due to their

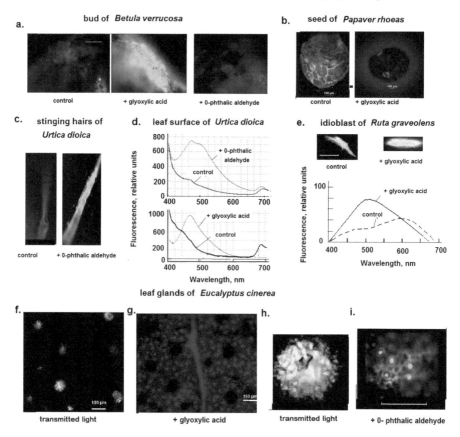

FIGURE 4.3 Fluorescence of secretory cells in medicinal plants before and after the treatment with *o*-glyoxylic acid for dopamine or phthalic aldehyde for histamine, seen at UV light (360–405 nm) in (a) bud of *Betula verrucosa* (bar = 200 μm); (b) seed of *Papaver rhoeas* (bar = 100 μm); (c) and (d) stinging hairs of *Urtica dioica* (bar = 50μm) on leaves and leaf fluorescence spectra received by spectrofluorimetry; (e) fluorescent image of root idioblast of *Ruta graveolens* (bar = 20 μm) and the fluorescence spectra recorded by microspectrofluorimetry; (f and g) the leaf surface of *Eucalyptus cinerea* with gland and its individual gland.

content of flavonoids and ester oils as active components. Staining with reagents for biogenic amines showed bright blue-green fluorescence with maxima at 460–500 nm on treatment with glyoxylic acid, which demonstrated high levels of dopamine in material collected in February and March. Green-yellow fluorescence allows the preparation to be visualized without cutting for slides. This period is characterized by the beginning of growth, when dopamine plays the role of a regulator of plant development (Roshchina 1991, 2001; Kulma and Szopa 2007). As for the staining by *o*-phthalic aldehyde for histamine, the reaction in blue was weak in a comparison with a control. Unlike *Betula* buds, stinging hairs (emergences) of *Urtica dioica* demonstrated high histamine concentration according to the emission in blue-yellow. Leaves of *Urtica dioica* with stinging hairs are collected in the blossom period and used in tinctures to treat hemorrhoidal, nasal, lung, uterine, and intestinal bleeding (Golovkin et al. 2001). The fluorescence spectra recorded with spectrofluorimeter demonstrated the presence of both histamines and catecholamines (Figure 4.3d). In fluorescent analysis for biogenic amines, besides the observation of samples under luminescence microscope (images b, c, e–i), microspectrofluorimetry was also used (image e). Unlike the above-mentioned species, the oil-containing species *Eucalyptus cinerea* grown in a Sochi dendrarium has no autofluorescent glands, and they also do not fluoresce after staining with *o*-glyoxylic acid (Figure 4.3f and g). The structures appear as dark spots, but the surrounding nonsecretory tissue structures emit in blue, showing the presence of dopamine. Individual glands appear in transmitted light as yellow (image h) and after staining with *o*-phthalic aldehyde, fluoresce in blue (image i). This shows the presence of histamine in the secretory structure. The leaves of many species belonging to the genus *Eucalyptus* (family Myrtaceae) demonstrate medicinal properties against viral and bacterial infections due to many monoterpenes and sesquiterpenes in the essential oil (Golovkin et al. 2001). The roots of another medicinal plant, *Ruta graveolens*, contains alkaloids in whole plants and may be used as an abortive, anti-inflammatory, antihistamine, and antispasmolytic and to aid in childbirth as well as having antitumor activity (Rastitelnie Resursy of USSR, 1985–1993; Golovkin et al. 2001). In particular, this plant contain alkaloids arborinine, gravacrydondiol, dictamine, kokusagenine, platydesmine, ribaline, rutacridone, rutalin, rutaverine, coumarins isoimperatorin, bergenin, xanthyletin, marmesin, and rutamarine, many furanocoumarins, including psoralen, valerinic acid, terpenoid daphnoretin, camphora, flavonoids quercetin and rutin as well as methylsalicylate (Golovkin et al. 2001). In the fluorescence spectra (Figure 4.3e), the secretory cell idioblast emits in orange-red (maxima 475 and 620 nm), and after staining with glyoxylic acid, it becomes green-yellow with a maximum 520 nm. This shows the presence of dopamine in the secretory cell.

It should be noted that catecholamines are predecessors of many important medicinal pharmaceuticals, such as alkaloids. An obvious example is metabolism in the *Papaver* genus (family Papaveraceae). Thus, in Figure 4.3b, one can see catecholamines in excretions from seeds of *Papaver rhoeas* Linn that fluoresce in blue. According to Sarin (2003), callus culture of the species showed the presence of three opium alkaloids: morphine, thebaine, and narcotine. This plant is also used in medicine, especially due to its analgesic and anti-inflammatory properties (Hajiai et al. 2018) as well as in folk medicine for the treatment of inflammation, cough, diarrhea, respiratory problems, asthma, insomnia and pain (Oh et al.2018). It is a traditional medicinal plant of Morocco, although it contains lower levels of opium products than *Papaver somniferum*. The use of fluorescence when isolated organelles from plant cells were stained histochemically with reagents to catecholamines or histamine, as well as emission after treatment with antagonists, showed that intracellular neurotransmitters can bind to certain cellular compartments. After the treatment with glyoxylic acid (a reagent for the determination of catecholamines), a slight blue emission with a maximum 460–475 nm was seen in nuclei and chloroplasts (in control variants in this spectral region, noticeable emission was absent) and very intense fluorescence (more than 20 times as compared with control) in the vacuoles (Roshchina et al. 2015; Roshchina 2016b, 2019a). After exposure to *o*-phthalic aldehyde, blue emission was more noticeable in nuclei and chloroplasts, which correlates with the previously observed effects in intact cells such as pollen and vegetative microspores. Among various organelles, the comparison of the emission intensity related to the biogenic amines showed the greatest effect in vacuoles, while chloroplasts have the weakest. Thus, on the surface, and possibly within the organelle, the fluorescence may demonstrate the presence of biogenic amines.

Dopamine and histamine of plants can be excreted out cells (Roshchina 2019a, b, c). Unlike animal cells, where catecholamines have been found in secretory vesicles and in the mitochondrial fraction, in plants, the compounds are shown to be present in vacuoles of *Papaver bracteatum* (Kutchan et al. 1988) as well as in isolated chloroplasts from some leguminous plants and *Urtica dioica* (Roshchina 2014, 2016; Roshchina et al. 2016a). As for histamine localization in the plant cell, in the literature, it is thought to be present only in vacuoles and/or secretory vesicles, whereas in animal cells, the neuromediator is also found in nuclei, mitochondria, and microsomes (Roshchina 1991).

4.1.2.3 Concentration of Biogenic Amines

The concentration of biogenic amines was first estimated in samples where it was unlikely to be present in large quantities. This was done by making concentration curves for dried films of biogenic amines treated with molecular probes (Roshchina and Yashin 2014; Roshchina 2014; Roshchina et al. 2015). Excitation at 360 nm was optimal. Moreover, actinic light at 405–430 nm was also suitable for the films if the emission was green or green-yellow. Low concentrations of noradrenaline were less well determined, unlike dopamine, which was the basis for catecholamine estimation. The concentration curves for films were used to determine the compounds.

The emission spectra have been recorded by various types of spectrofluorimeter or the microspectrofluorimeter MSF-2 depending on the object of the study. The registration of fluorescence intensity was expressed in relative units in three to four samples per variant. The results underwent statistical analysis. The mean $\pm$ standard error (SEM) was shown in table data or graphically on histograms, and the relative standard deviation (RSD) was also estimated.

For an approximate estimation of the amount of biogenic amines in individual cells, dried films of dopamine or histamine in different concentrations from 10^{-8} to 10^{-2} M on slides were prepared. The emission intensity (I) of intact cells excited by UV (360–380 nm/360–405 nm) or violet light (420–435 nm) of intact cells was recorded in the spectral region at 450–460 nm (I_{450}) or at 520–550 (I_{530}) for samples with high concentrations of the amines.

The above-mentioned assays were performed for pollen from 25 species as well as leaf and stem stinging hairs of *Urtica dioica* and vegetative microspores of horsetail, *Equisetum arvense* (Roshchina and Yashin 2014; Roshchina 2014; Roshchina et al. 2015). The cells and their excretions excited by UV light may fluoresce in blue-green after staining with glyoxylic acid or in blue/blue-green or yellow after treatment with *o*-phthalic aldehyde. The reagents penetrated into the cells and stained DNA-containing organelles such as nuclei and chloroplasts. The attempt to estimate the occurrence and content of catecholamines and histamine by the fluorescent method in individual plant cells may be useful for express diagnostics of these biogenic amines as biologically active compounds in ecosystems and drugs in medicinal plants (Roshchina et al. 2015; Roshchina 2016a, b). This method is rarely used for serotonin, because high-performance chromatography is more often applied (Mukherjee and Ramakrishna 2019).

As shown in some publications (Roshchina and Yashin 2014; Roshchina 2014; Roshchina et al. 2015), the fluorescent method is able to determine biogenic amines in pollen grains. Among 24 plant species studied, increased amounts of catecholamines and/or histamine were observed in pollen of *Acer platanoides* L., *Anemone ranunculoides* L., *Anthriscus sylvestris* L., *Betula verrucosa* Ehrh., *Corylus avellana* L., *Larix decidua* Mill., *Hippeastrum hybridum* L., *Narcissus pseudonarcissus* L., *Picea excelsior (excelsa)* L., *Pinus sylvestris* L., *Populus balsamifera* L., *Syringa vulgaris* L., *Taraxacum officinale* L., *Tilia cordata* Mill., *Tussilago farfara* L., and *Tulipa* sp. This fact is related to the peculiarities of nitrogen metabolism of the plant species. Moreover, most fluorescent data confirm the experiments with similar pollen species done by Marquardt and Vogg (1952) with the help of pharmacological and chemical methods. This information is very valuable for medicine, because it provides knowledge about allergic factors (dopamine and histamine content). We should also remember about possible allergic effects of pollen, rich in the biogenic amines, as admixture in fresh raw pharmaceutic material, or when the microspores are used as natural drugs in medicinal care.

Express diagnostics of neurotransmitters such as catecholamines and histamine in single plant cells by a fluorescent method is needed for monitoring of metabolic processes under various environmental factors and may apply in medicine and pharmacology. The investigations are necessary to understand the effects induced by natural drugs and the nutritional additions contained in plant cells or their extracts. The reactions may be carried out in the following ways. (1) The accumulation of catecholamines in plants is one indicator of their development and sometimes, of stress reactions (Kamo and Mahlberg 1984; Swiedrych et al. 2004). (2) Compounds known as regulators of metabolism, growth, and development in any living cell may therefore play a role in chemical relations between different organisms in an ecosystem (Roshchina 2010, 2019c, Roshchina et al. 2019). (3) Histamine also participates in the recognition of pollen from its own plant species by the pistil stigma (Roshchina and Melnikova 1998a, b). If we keep in mind that histamine is a component of the human allergic reaction to the pollen of different plant species and occurs in air and water in certain periods (Stanley and Linskens 1974), experiments in this field become relevant to clinical practice (Heidinder et al. 2017), when the knowledge is used in ophthalmology. (4) Experiments on single intact cells or isolated organelles as model systems (Roshchina 2004, 2007b) have potential. For example, vacuoles of the African folk medicinal plant *Saintpaulia ionantha* have a characteristic blue fluorescence only on the outer membrane after staining with *o*-phthalic aldehyde for histamine. Penetration into the organelles occurs for the vacuoles of *Euphorbia milii*. Here, blue-green emission also appears inside the organelles after the addition of glyoxylic acid, as seen in optical slices under a confocal microscope (Roshchina et al. 2018). In this connection, there is a very relevant question about the form of biogenic amines within plant cells. One monograph (Roshchina 2001) contains information about the presence of dopamine in the form of glycosides in seeds of *Entada pursaetha* and the occurrence of histamine (depending on the species of sample and organ) in the form of N-acetylhistamine, N,N-dimethylhistamine, cinnamoylhistamine, or feruloylhistamine as well.

The presence of biogenic amines may be estimated using agonists and antagonists of neurotransmitters fluorescent in blue and blue-green and bound to cellular membranes. Figure 4.4. shows some examples of binding of glyoxylic acid and its conjugate as well as an antagonist of serotonin, inmecarb. In the case of serotonin,

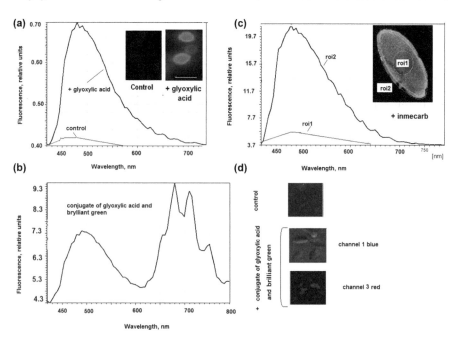

FIGURE 4.4 Fluorescence of pollen from medicinal plant *Clivia miniata* Regel (bar = 50 μm), after histochemical staining with glyoxylic acid for dopamine (a) and with inmecarb, antagonist of serotonin (b). The fluorescence spectra were received by laser-scanning confocal microscopy, laser excitation 405 nm. Images seen under Leica DM 6000 B luminescence microscope (a) and with Leica TSC SP 5 laser-scanning confocal microscope (b and c). ROI: parts from which the fluorescence spectra were received (ROI 1 from cellular interior and ROI 2 from the surface).

besides formaldehyde acting also on catecholamines, we have no tools and may use a fluorescent antagonist as a molecular probe. Various assays for the staining of dopamine with glyoxylic acid use conjugates of the acid with brilliant green in order to show where the reagents bind, as in the example of pollen from the medicinal plant *Clivia miniata* (Figure 4.4a, b). In the first case, we see blue emission on the surface of the single cell (a). The conjugate of glyoxylic acid with brilliant green (b) acted on pollen and was analyzed by scanning confocal microscopy. Figure 4.4d confirms this in color: blue in the blue channel and red in the red channel of the confocal microscope. Beside formaldehyde as histochemical reagent for serotonin, it is possible to use also fluorescent antagonist of the biogenic amine –inmecarb – which binds with the serotonin receptor in plant cells and fluoresces in blue (Roshchina 2017). After the treatment of *Hippeastrum hybridum* pollen with inmecarb, in the emission spectrum there was maximum 460–465 nm recorded by microspectrofluorimeter MSF-1. On the model of *Clivia miniata* pollen moisted with the solution of the antagonist, one can see green fluorescence registered by laser-scanning confocal microscopy, and in the spectrum maximum was 475 nm (Figure 4.4c). The green emission has been located on the pollen surface. ROI 1 of the internal part of pollen shows that inmecarb penetrates into the cell, although most of the reagent bound on the surface, where the serotonin receptors are present.

4.1.2.4 Staining of Cells with Fluorescent Analogs of Biogenic Amines and Their Antagonists

Fluorescent antagonists binding with appropriate receptors have been described for model single cells such as pollen of various species and vegetative microspores of *Equisetum arvense*. In the plant cells, as antagonists following compounds were used : d- tubocurarine – for receptor of acetylcholine (cholinoreceptors of the nicotinic type), yohimbine – for receptors of dopamine and norepinephrine known as adrenoreceptors (Roshchina 2008, 2014; Roshchina and Karnaukhov 2010), inmecarb - for the serotonin receptors (Roshchina 2014, 2017, 2019a, b) and azulene - for the histamine receptors (Roshchina and Karnaukhov 2010; Roshchina 2014). Intensity of the fluorescence is dependent on the sample. Binding of d-tubocurarine, yohimbine, inmecarb and azulenes was observed in whole cells (Roshchina 2008, 2014; Roshchina and Karnaukhov 2010) and some isolated organelles nuclei, chloroplasts and vacuoles (Roshchina 2016a, b, 2019b). A distinct increase in the emission throughout blue and blue-green spectral region (460–520 nm). Fluorescent reagents were bound on the surface, and the example for binding of the serotonin antagonist inmecarb with the surface of the pollen has been demonstrated in Section 4.4.2.3. Although there is possibility of the penetration of fluorescent antagonists into the cell. Sensitivity of cellular organelles to the compounds differed among species studied (Roshchina et al., 2016a). Increased blue-green emission after the treatment with inmecarb observed in the vacuoles of *Beta vulgaris*, but not those of *Euphorbia milii*. As seen from the fluorescence measurement, chloroplasts isolated from leaves of *Euphorbia milii* may bind d-tubocurarine and yohimbine, but not – inmecarb. Nuclei, isolated from petals of *Kalanchoe blossfeldiana*, fluoresced, only binding with yohimbine (Roshchina et al. 2016a, Roshchina 2016b).Another approach to histochemical analysis consists in using fluorescent-labelled analogs of neurotransmitters, serving as their agonists. These are Bodipy-dopamine or Bodipy-serotonin (Roshchina et al. 2003). Earlier in order to define sites of binding with the agonists in the pollen cells from *Hippeastrum hybridum* or vegetative microspores from *Equisetum arvense* (Roshchina 2016a, b) were used. Then the structural images of fluorescent objects received by laser-scanning confocal microscopy, and the fluorescence spectra of whole cells were registered by microspectrofluorimetry. The image of optical slices of nuclei isolated from petals of *Philadelphus grandiflorus* after treatment with Bodipy-dopamine (Bezuglov et al. 2004) demonstrated the agonist emission only on the periphery of the organelle (Roshchina 2006a, b; Roshchina et al. 2016a, b). Among other fluorescent synthetic analogs, FITC-DA (FITC-dopamine), FITC-L-NE (FITC-noradrenaline, norepinephrine) and FITC-5-HT (FITC-serotonin) have been represented by Lin with co-workers (2015), but not tested on plant cell yet. In Figure 4.4c, we can see how the pollen of *Clivia miniata* fluoresces after staining with the fluorescent antagonist of serotonin, inmecarb. The green emission takes place on the surface of the microspore.

4.1.2.5   Biogenic Amines as Stress Compounds in Plant Material

Histochemical studies of pollen and vegetative spores as unicellular model systems for the determination of catecholamines and histamine (with glyoxylic acid and *o*-phthalic aldehyde, respectively) have shown that ozone in various concentrations influences their accumulation in plant cells (Roshchina and Yashin 2014; Roshchina et al. 2015a, b; Roshchina 2016b). Enhanced amounts of the neurochemicals in pharmaceutic materials show that the plant has lived and grown under stress. By microspectrofluorimetry and laser-scanning confocal microscopy, these stress metabolites – neurotransmitters – were detected by their characteristic fluorescence at 460–480 nm in the samples, and increased concentrations of catecholamines and histamine were observed under treatment with ozone, as assessed by an increase in the intensity of the emission. After 6 hours' exposure (total dose of 0.012 μL/L) to ozone, not only did the level of catecholamines increase in plants important for pharmacy, but in the pollen of poplar, *Populus balsamifera*, the histamine content significantly increased too. Higher concentrations of catecholamines in comparison with control were found in vegetative microspores of *Equisetum arvense* and pollen of *Corylus avellana* and *P. balsamifera* after exposure to large doses of ozone (0.032 μL/L). Samples from *P. balsamifera* were especially sensitive; in comparison with a control, the concentrations of dopamine increased more than threefold and those of histamine 2.5-fold (Roshchina et al. 2015a, b). These neurochemicals' reactions were noted both in secretions and in some organelles within cells (nuclei and chloroplasts). The use of fluorescent analysis on cells as bioindicators of ozone may be useful in ecomonitoring for early warning of concentrations of this compound in the air that are hazardous to health.

Analogous histochemical analysis for biogenic amines has been carried out in conditions where single cells of pollen and vegetative microspores were grown in high salt concentrations (Roshchina and Yashin 2014). The enhanced fluorescence intensity of *Equisetum arvense* microspores from sporophytes in histochemical reaction with *o*-phthalic aldehyde was studied before and after 24 h of germination in Na_2SO_4 solutions. Secretions from the cell emitted especially brightly at the 1%

salt concentration, which shows an increased concentration of histamine released. Dependence of histamine accumulation in seeds of sunflower *Helianthus annuus* was found by Korobova with coworkers (1988).

The microbiota produces neurochemicals that exert an important influence on the host plant organism, both physiological and morphogenetic (Lyte and Ernst 1992; Lyte 2010). A large number of neurochemicals, including biogenic amines such as catecholamines, serotonin, and histamine, are formed by microorganisms, and their increased amounts may be analyzed for histochemical reactions in plants (Oleskin and Shenderov 2019).

In some cases, catecholamines participate in plant tolerance to drought stress (Gao et al. 2019). Thus, seed pretreatment with 100 μM dopamine alleviated drought stress in apple seedlings (*Malus domestica* Borkh.). Dopamine inhibited the degradation of photosynthetic pigments and increased the net photosynthetic rate under drought stress (Gao et al. 2018). It may improve apple drought tolerance by activating Ca^{2+} signaling pathways through increased depression of many genes. In model experiments on isolated chloroplasts of *Pisum sativum*, dopamine in pair with ascorbic acid donated electrons into the electron transport chain at the level of Photosystem I (Roshchina 1990b).

The location of catecholamines in plant cells has been studied in some species. Dopamine has been found in vacuoles of cells of poppy *Papaver bracteatum* and *P. somniferum* (Kutchan et al. 1986). Noradrenaline and adrenaline occur, mainly, in vacuoles and have also been identified in isolated chloroplasts of some plants of the families Fabaceae and Urticaceae (Roshchina 1991, 2001).

The presence and localization of the neurotransmitters catecholamines and histamine within single plant cells have been studied by V.V. Roshchina and coauthors (2014, 2015a, b) in pollens by fluorescence microscopy and its modification as microspectrofluorimetry using special molecular probes—glyoxylic acid (reagent for catecholamines) and *o*-phthalic aldehyde (reagent for histamine) in various conditions with the natural oxidant ozone. Among 25 species, biogenic amines accumulated in pollen grains from wind-pollinated medicinal plants such as *Betula verrucosa*, *Corylus avellana*, and *Populus balsamifera*. Fluorescence related to catecholamines was found, mainly, in *C. avellana*, while emission related to catecholamines and histamine was only seen in microspores *of P. balsamifera*.

Histamine poisoning is the most common food-borne problem caused by biogenic amines. The allergic reaction is characterized by difficult breathing, vomiting, rash, fever, diarrhea, hypotension, headache, and pruritus (Fernández et al. 2006). Scombroid poisoning (toxicosis by histamine-like products) is an important food-borne disease. High concentrations of dopamine also show damage (Nalbandyan 1986) due to binding with copper-containing proteins of the brain, which leads to schizophrenia. Some complexes were studied by Walaas et al. (1963). Another example relates to links with actin of muscles; it may block normal contraction, as shown in an animal model (Shubina et al. 2011).

The accumulation of catecholamine and histamine concentrations often leads to stress. In this case, a fluorescent method may help in express diagnostics of biogenic amines in cells. Kucharski (2005) used a fluorescence method for the identification of weeds resistant to the inhibitors of photosynthesis, such as the herbicides linuron, bentazone, chlortoluron, and isoproturon, and showed that the medicinal plant poppy *Papaver rhoeas* accumulates alkaloids originating from catecholamines (Hajjaj et al. 2018; Oh et al. 2018). The latex in the seedpods is narcotic and slightly sedative; it needs to be used in very small quantities under expert supervision as a sleep-inducing drug. The leaves and seeds are tonic. They are useful in the treatment of low fevers. The plant also has anticancer properties. Toxicosis for mammals is possible, though the toxicity is low. The flowers and petals, harvested as the flowers open, are dried for later use as anodyne, emollient, emmenagogue, expectorant, hypnotic, slightly narcotic, and sedative. An infusion of the plant has been applied internally in the treatment of bronchial complaints and coughs, insomnia, poor digestion, nervous digestive disorders, and minor painful conditions. The petals of the flowers are also used in the treatment of jaundice.

Plants also can perceive abiotic stresses and elicit appropriate responses with altered metabolism, growth, and development. Dopamine prevents many plant dysfunctions quickly in plants under drought conditions (Gao et al. 2019). For sunflowers, higher soil salt concentrations led to the formation of histamine in sunflower seeds (Korobova et al. 1988). Under salt stress, the biosynthesis of catecholamines and histamine in vegetative microspores of *Equisetum arvense* increased (Roshchina and Yashin 2014).

In nature, the collected leaves of some plants with valuable medicinal compounds in secretory structures may also contain dopamine and histamine, as seen in their images after staining with histochemical reagents (Table 4.3).

As seen in Table 4.3, many subtropical plant species from the Sochi region of Russia contain dopamine, but marked amounts are seen in the leaf surface hairs of *Actinidia chinensis* and *Ficus colchicium* as well as in leaf glands of *Ginkgo biloba*. The neurotransmitter is also present in flower glands of *Nerium oleander* and in seeds and secretions of *Papaver rhoeas*. Often, the surface around dark (non-fluorescent) glands contains dopamine, as in *Citrus unshu*. Around the terpene-containing leaf glands of *Eucalyptus cinerea*, which have no autofluorescence, blue emission was seen after staining with glyoxylic acid. Therefore, the emitting surface contains dopamine (catecholamines).

The accumulation of an increased amount of histamine (yellow fluorescence), specific to the surface around non-fluorescing terpene-rich glands, was observed in leaves of *Citrus unshu* (Table 4.3.). This is characteristic of stress conditions during plant growth. A marked amount of histamine is observed in some parts of the terpene-containing leaf glands of *Eucalyptus cinerea* and in leaf glands and stomata of *Buddleja officinalis*. Often, dopamine is absent, but histamine is found in glands, as in leaf of *Cistis ladaniferus* or seeds of *Papaver rhoeas*. In some medicinal plants, such as *Hamamelis* and *Cassia floribunda*, dopamine and histamine are absent in both leaves and flowers.

4.2 PHENOL OCCURRENCE IN PLANT SECRETORY CELLS

Phenols in plant secretory cells have been studied many times in various publications (Fahn 1979; Roshchina V.D and Roshchina V.V. 1989; de Castro and Demarco 2008). Besides the studies of autofluorescence of many phenolic compounds, there are

TABLE 4.3

Fluorescence of Secretory Structures after Histochemical Staining with Glyoxylic Acid and *o*-Phthalic Aldehyde for Determination of Dopamine and Histamine, Respectively

Species and Possible Active Matter	Organ	Control without Staining (Autofluorescence)	Fluorescence at 450–460 nm Staining with	
			Glyoxylic Acid	*o*-Phthalic Aldehyde
Actinidia chinensis Planch., monoterpenic alkaloid actinidine (roots)	Leaf (upper side)	Hairs bluish-yellow	Increased blue emission in blue hairs (dopamine)	Increased blue emission of surface (histamine)
	Leaf (lower side)	Hairs bluish-yellow	Increased blue emission in blue hairs (dopamine)	Increased blue emission of surface (histamine)
Albizia julibrussin Durazz, acetylcholine, noradrenaline, serotonin	Leaf (upper side)	Non-fluorescent glands(appear as dark spots)	Slight emission (small amount of dopamine)	Slight emission(small amount of histamine)
	Leaf (lower side)	Non-fluorescent glands(appear as dark spots)	Slight emission (small amount of dopamine)	Slight emission (small amount of histamine)
Buddleja officinalis Maxim, flavonoids quercetin, kaempferol, linarin	Leaf (upper side) Yellow glands in transmitted light	Gland with blue autofluorescence	No increase in blue emission	Increased blue fluorescence in glands and stomata (histamine)
	Leaf (lower side)	Gland with blue autofluorescence	No increase in blue emission	Increased blue fluorescence in glands and stomata (histamine)
Cassia floribunda L., cinnamaldehyde, flavones diosmetin, luteolin	Flower Leaf	Dark channels and glands— no emission	No emission	No emission
Cistus ladaniferus L., antiseptic, kaempferol, phenolcarbonic acids	Leaf (upper side)	No emission of glands	No emission of glands	Increased blue fluorescence in glands (histamine)
	Leaf (lower side)	No emission of glands	No emission of glands	Increased blue fluorescence in glands (histamine)

(Continued)

TABLE 4.3 (CONTINUED)

Fluorescence of Secretory Structures after Histochemical Staining with Glyoxylic Acid and o-Phthalic Aldehyde for Determination of Dopamine and Histamine, Respectively

Citrus unshu Marc., vitamins	Leaf (upper side); emitting glands (white spots) in transmitted light	Weak blue fluorescence of surface with dark glands	Weak common fluorescence with non-fluorescent glands	Yellow fluorescence around dark glands (histamine in high concentrations)
	Leaf (lower side); emitting glands (white spots) in transmitted light	Weak blue fluorescence of stomata around dark glands	Increased blue fluorescence in cells around dark glands (dopamine)	Yellow fluorescence around dark glands (histamine in high concentrations)
Eucalyptus cinerea F. Muell., flavonoids, monoterpenes, carvone	Leaf, yellow-colored glands in transmitted light	No emission	No gland fluorescence, but fluorescence around the gland (dopamine)	Blue fluorescence of some parts of gland(histamine)
Ficus carica L., flavonoids	Leaf (lower side), channels, hairs, a lot of dark glands	Blue emission of many glands and channels	Increased blue fluorescence in the base of hairs (dopamine)	Weak fluorescence
	Leaf (upper side), channels, dark glands, hairs	Blue emission of many glands and channels	Increased blue fluorescence in the base of hairs (dopamine)	Weak fluorescence
Ginkgo biloba L., anacardic acid, bilobalide, riboflavin, triterpenoids, ginsengosides	Leaf (upper side)	Dark glands, no emission	Blue emission (dopamine)	No fluorescence, no histamine
	Leaf (lower side)	Few glands, no emission	Blue emission (dopamine)	No fluorescence, no histamine
Hamamelis japonica Sieb. et Zucc, flavonoids quercetin, myricetin	Leaf glands seen as red in transmitted light	No fluorescence	No fluorescence	No fluorescence
Heimia salicifolia, alkaloids	Flower petal	No fluorescence	Blue fluorescence (dopamine)	Weak emission (small amount of histamine if any)

(Continued)

TABLE 4.3 (CONTINUED)

Fluorescence of Secretory Structures after Histochemical Staining with Glyoxylic Acid and o-Phthalic Aldehyde for Determination of Dopamine and Histamine, Respectively

	Leaf	Weak fluorescence	Blue fluorescence (dopamine)	Weak emission (small amount of histamine if any)
Nerium oleander L., anticancer adinerin, cardenolides karabin	Flower	Blue fluorescent glands	Blue fluorescence (dopamine)	No emission
	Leaf	Dark glands (no emission)		
Papaver rhoeas L., alkaloids, dopamine, aloe-emodin (anthraquinones)	Seeds(brownish in transmitted light)	No fluorescence	Blue fluorescence of surface and secretions (dopamine)	No fluorescence
	Fruit skin	No fluorescence	No fluorescence	No fluorescence

Sources: Golovkin et al. 2001; Rumala, C.S., et al., *Phytochemistry*, 69, 1756–1762, 2008; Roshchina, V.V., *Biomediators in Plants. Acetylcholine and Biogenic Amines*, Biological Center of USSR Academy of Sciences, Pushchino, 1991; Roshchina, V.V., *Neurotransmitters in Plant Life*, Science Publishers, Enfield, NH, 2001.

histochemical methods to determine these substances, if necessary, by fluorescence microscopy. This method enables the visualization of phenol location in secretory structures using fluorescent markers. Muravnik (2008) described the application of the fluorochrome β-aminodiethyl ester of diphenylboric acid according to Heinrich with coauthors (2002) as a marker for the presence of flavonoids on the heads of trichomes of various species of the genus *Juglans* (many of them used in the pharmaceutical industry). Fluorescence was observed in blue after excitation by UV light. This method has also been used for the medicinal plant *Tussilago farfara* (Muravnik et al. 2016) for the analysis of phenols in capitate glandular trichomes.

Histochemical investigations with natural product reagents for phenol determination have been described by Combrinck with coworkers (2007) for the medicinal species *Lippia scaberrima*, an aromatic indigenous South African plant with medicinal applications. The plant is an aromatic, perennial shrub, which reaches a height of about 60 cm and prefers to grow in a dry climate with summer rainfall. Medicinal tea made from the leaves of *L. scaberrima* has been sold commercially as a general health tea under the name *Mosukujane tea*. The activity of medicinal plants can be attribute to their secondary metabolites, some of which occur in the essential oils. The above-mentioned authors prepared a mixture of aqueous $AlCl_3$ solution (5%) and 0.05% diphenylboric acid-b-ethylaminoester in 10% methanol, as described by Heinrich et al. (2002) for the detection of flavonoids. Entire leaves were soaked and dried on adsorbent paper, cut in half across the mid-vein, and mounted in oil, both adaxial and abaxial sides facing upwards. Vanillin-HCl was prepared by dissolving 0.1 g vanillin in 100 mL concentrated HCl (Nikolakaki and Christodoulakis 2004). This reagent is a universal indicator of various compounds, enabling differentiation between different classes of secondary metabolites such as flavonoids, terpenoids, and phenyl carboxylic acids. For examination of autofluorescence and staining, leaves were mounted on a glass slide, without a cover glass, and viewed directly under UV light (excitation 365 nm; emission 397 nm). A blue filter was used to indicate further fluorescence (excitation 386 nm). In works devoted to phytopathology (Valette et al. 1998), flavonoid compounds were detected with Neu's reagent. Sections were immersed for 2 to 5 min in 1% 2-amino-ethyldiphenylborinate (Fluka, Buchs, Switzerland) in absolute methanol, mounted, and examined by epifluorescence microscopy. Flavonoids were stained yellow-orange under UV light and yellow in lemon-glycerin-water when observed with blue light. After treatment of sections with Neu's reagent, caffeic esters were stained blue-white and ferulic acid stained blue when UV illuminated. The characterization of dopamine (whitish blue under UV light) was performed on sections immersed in a solution of lactic and glyoxylic acids and heated at 100 °C for 1 min (Andary et al. 1996). Safranine, an azo dye, is commonly used for plant microscopy, especially as a stain for lignified tissues, such as xylem. This dye fluorescently labels the wood cell wall, producing green/yellow fluorescence in the secondary cell wall and red/orange fluorescence in the middle lamella region (Bond et al. 2008). The fluorescence behavior of safranine under blue light excitation was studied using a variety of wood- and fiber-based samples of known composition to interpret the observed color differentiation of different cell wall types. Bond with coworkers (2008) examined the spectra using spectral confocal microscopy for differences in the fluorescence emission between lignin-rich and cellulose-rich cell walls, including damaged wood and decayed wood compared with normal wood in some samples of *Pinus radiata* D. Don (radiata pine), *Pseudotsuga menziesii*, and *Populus deltoides* Bartr. ex Marsh (poplar). The results indicated that lignin-rich cell walls, such as the middle lamella tracheids, the secondary wall of compression wood tracheids, and wood decayed by brown rot, tend to fluoresce red or orange, while cellulose-rich cell walls, such as resin canals, wood decayed by white rot, cotton fibers, and the G-layer of tension wood fibers tend to fluoresce green/yellow. This variation in fluorescence emission seems to be due to factors including an emission shift toward red wavelengths combined with dye quenching at shorter wavelengths in regions with high lignin content. Safranine fluorescence provides a useful way to differentiate lignin-rich and cellulose-rich cell walls without counterstaining as required for bright field microscopy.

4.3 TERPENOID OCCURRENCE IN PLANT SECRETORY CELLS

Among fluorescent histochemical methods, there are only a few real dyes for lipids that have been applied to plant tissues. Among these is auramine O, used for cut sections of leaves from medicinal *Agave* spp. as well as for *Dianthus* and *Brassica* spp. (Gahan 1984). The samples are stained in a 0.01% solution of the dye for10 min, after which unsaturated acidic waxes fluoresce in greenish-yellow under blue light excitation in a fluorescence microscope. In some cases, many fluorescent flavonoids and coumarins, contained in secretory cells enriched oils, serve as markers for lipids and lipophilic compounds in thin-layer chromatography (Wagner and Bladt 1996).

4.4 NUCLEIC ACIDS AS AN INDICATOR OF THE VIABILITY OF FRESH RAW MATERIAL

With various dyes, a researcher may determine the state of nucleic acids using commercial reagents: acridine orange, Hoechst 33342, ethidium bromide, or some azulenes/proazulenes. The main formulae are shown in Figure 4.5. This enables the researcher to know the viability of cells in phytomaterials and cellular damage due to unfavorable growth conditions of a medicinal species; for example, apoptosis and possible mutagenesis. This information is valuable for the selection and cultivation of medicinal plants as well.

Some natural compounds, such as azulenes, proazulenes, anthocyanins, and alkaloids, are also capable of binding to nucleic acids, so that they can be recommended for the same purpose. In Table 4.4, several recommended fluorescent dyes and the spectral region of their emission on excitation by light at 360–380 nm are presented.

4.4.1 ARTIFICIAL DYES

Dyes (10^{-6}–10^{-4} M depending on the objects being analyzed) penetrate the cell and bind to nucleic acids. When these cells are excited with UV or violet light, fluorescence is induced, which is better seen in the DNA-containing organelles nuclei and chloroplasts. Hoechst-33342 and ethidium bromide reagents may be used to see alterations in chromatin structure due to the degradation of proteins and nucleic

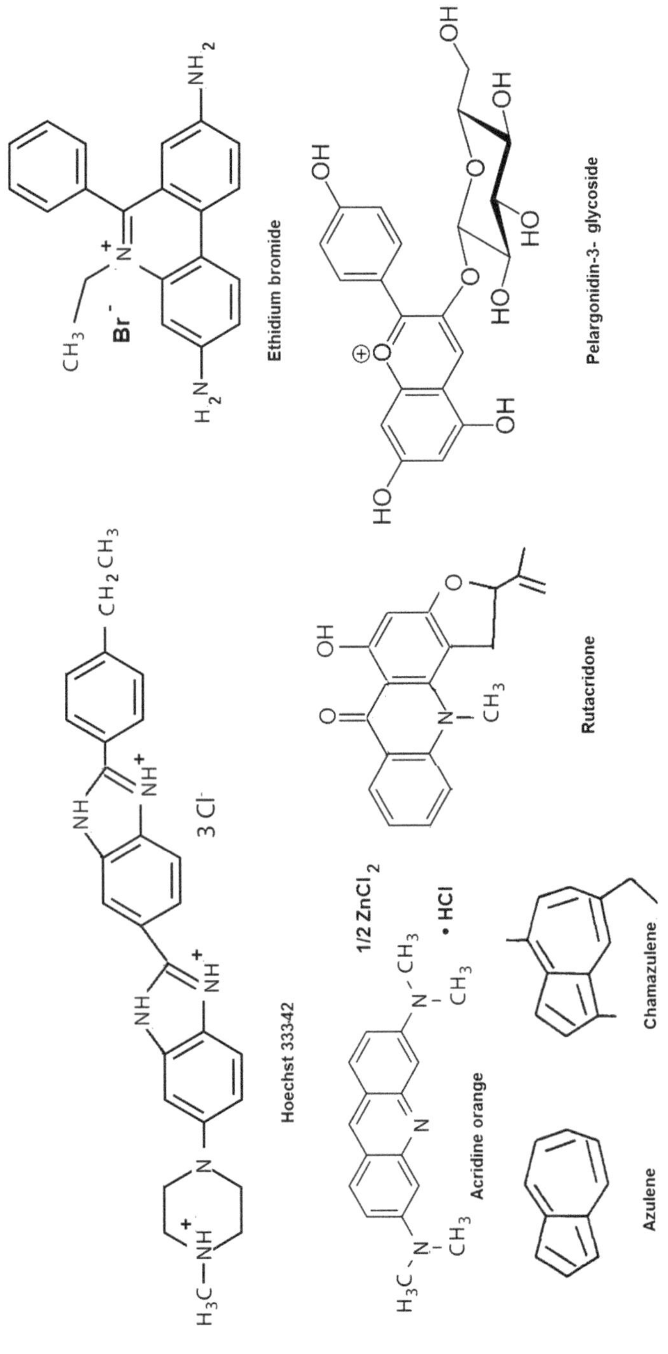

FIGURE 4.5   Formulae of dyes for nucleic acids.

TABLE 4.4
Dyes for Nucleic Acids in Plant Cells

Reagent	Color	Fluorescence Maximum, nm	Target in Nucleus
Acridine orange	Green	520–530 when it binds to DNA	Interacts with DNA and RNA by intercalation or electrostatic attractions
	Red	650 when it binds to RNA	
Azulene	Blue	430	Increased the fluorescence of guanosine in nucleic acids
Ethidium bromide	Orange	605	Fluorescence increases 21-fold upon binding to double-stranded RNA and 25-fold on binding to double-stranded DNA
Gaillardine (proazulene)	Blue	430	
Hoechst 33342	Blue	460–461	Binds to the minor groove of double-stranded DNA with a preference for sequences rich in adenine and thymine
Pelargonidin	Blue or blue-green	450, shoulder 520	Proposed to be the phosphorus between two deoxyribose residues of the DNA near the contact of ribose (deoxyribose) with nucleotides
Rutacridone	Green	520–530 nm	Proposed links with desoxyribose in nucleic acids

Sources: Haugland, R.P., *Handbook of Fluorescent Probes and Research Products*, Molecular Probes, Leiden, 2001; Jaldappagari, S., et al., *Top. Heterocycl. Chem.*, 15, 49–65, 2008; Roshchina, V.V., *J. Fluoresc.*, 12, 241–243, 2002; Roshchina, V.V., *Allelopathy J.*, 13, 3–16, 2004; 2007a; Roshchina, V.V., *Fluorescing World of Plant Secreting Cells*, Science Publishers, Enfield, NH, 2008; Slaninová, Slanina J, Táborská, 2008; Chemicki Listy, 102: 427–433.

acids. In order to distinguish living cells from late apoptotic cells, methods of histochemical staining with fluorescent dyes have been developed (Haugland 2001). The dye Hoechst binds to t-RNA extremely effectively too (Vekshin 2011). The lifetime of Hoechst 33342 is enhanced from 0.4 to 4.4 ns on binding. It has an affinity for various polynucleotide chains. The main histochemical use of the dyes is assay of early apoptosis by staining DNA with Hoechst 33342, while late apoptotic cells have membranes permeable to ethidium bromide. The strengthened fluorescence of the DNA–Hoechst 33342 complex is characteristic of chromatin damage (Roshchina et al 1998b, c). An increased number of cells permeable to ethidium bromide indicates membrane damage. When pollen is exposed to the highest concentrations of ozone, for instance, cells becomes non-viable, and apoptosis appears to be a frequent occurrence (Roshchina V.V and Roshchina V.D. 2003). This is seen as increased fluorescence intensity of the cells. Similar damage to pollen cells of *Hippeastrum hybridum* under a high concentration of ozone (1–2.4 μL/L) in air was demonstrated as an increase in DNA–Hoechst 33343 complex fluorescence at 460 nm (Roshchina V.V.

and Roshchina V.D. 2003). Under this factor, there was the 100% staining of cells by ethidium bromide, which penetrates within pollen grains and, when the reagent achieves nucleus, it fluoresces (Roshchina V.V and Roshchina V.D. 2003). The diagnostics of cell damage concentrates on apoptosis (programmed cell death, which is the genetically controlled ablation of cells during development). This is particularly well illustrated by cancer cells or cells subject to genetic transformation. The form of apoptosis induced by ozone consists of hypersensitive cell death (Rao et al. 2000, Rao and Davis 2001). Moreover, O_3 has been considered as a tool for probing the processes involved in programmed cell death in plants (Rao et al. 2000). Apoptosis leads to the appearance of lesions and necrosis in plant leaves (Rao and Davis 1999). Special attention is always paid to the damaged plant cells that occur on leaf or pollen surfaces when exposed to chronically low, medium, or acutely high doses of ozone. Among such injuries are apoptosis (which requires special microscopic analysis) and visual types of damage (chlorosis, dots or lesions, pigmentation, and necrosis). The appearance of such damage is used to define the sensitivity of a whole plant species to ozone. Unlike necrotic cells, apoptotic cells have characterized by (1) relative membrane integrity and (2) alteration in chromatin structure due to the degradation of proteins and nucleic acids. In order to distinguish living cells from late apoptotic cells, methods of histochemical staining with fluorescent dyes have been developed (Haugland 2001). Results have been obtained by staining ozone-treated pollen with the above-mentioned dyes. According to Roshchina and Roshchina (2003), a general reduction in populations of the medicinal species plantain, *Plantago major*, has been observed near an area in Pushchino where automobiles stop and where tropospheric ozone appears. Fluorescent probes with Hoechst 33342 and ethidium bromide allow damage to be visualized before visible alterations on the leaves.

Viability of the cells in phytomaterials for pharmacy may also be observed with fluorescent reagents. In Figure 4.6a, the reader can see the presence of living nuclei

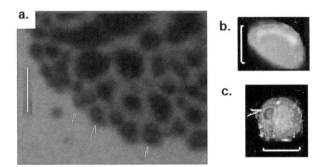

FIGURE 4.6   Images of cells after staining with 10^{-5} M reagents for DNA (nuclei shown with arrows). (a) Multicellular gland of female cone from *Humulus lupulus* treated with Hoechst 3342 under luminescence microscope (bar = 50 µm); (b) and (c) fluorescing images of the pollen from *Hippeastrum hybridum* under luminescence microscope (bar = 50 µm) and vegetative microspores from *Equisetum arvense* under Carl Zeiss laser-scanning confocal microscope (bar = 20 µm) treated with rutacridone. Sources: unpublished data and modified fragment of photo Roshchina et al., *Allelopathy J.* 28, 1–12, 2011.)

fluorescing in blue in an example of a multicellular gland from the female cone of *Humulus lupulus* stained with Hoechst 33342. Without such a procedure, we cannot see the organelles in the materials.

4.4.2 Natural Dyes

Fluorescent natural chemicals could be used as fluorescent dyes for the study of interactions between cells and analysis of cellular damage. Fluorescence within cells linked to nucleic acids (Table 4.5) has been observed for sesquiterpene lactones (proazulenes and azulenes) and alkaloids (Roshchina 2008). After similar histochemical staining, nucleus and chloroplasts fluoresce, unlike other cytoplasm. Pigmented secretions enriched in red anthocyanins (glycosides of pelargonidin from *Hippeastrum hybridum* petals), blue azulenes (synthetic azulene and azulene extracted from leaves of *Artemisia absinthium*), and yellow alkaloids (rutacridone from *Ruta graveolens* roots) also serve as tools for nucleic acid (Roshchina et al. 2011). More information about the fluorescence maxima of several natural dyes tested in various plant samples is in Table 4.5 (Roshchina 2008; Roshchina et al. 2011). All the compounds were tested on single cells of spores belonging to horsetail (*Equisetum arvense*) and pollen of *Hippeastrum hybridum* as well as on nuclei isolated from the soft non-pigmented tissues of petals of matricaria (wild camomile), *Matricaria reticuta* L., and colored by the above-mentioned reagents.

TABLE 4.5
Fluorescence of Some Individual Drugs Bound to Nuclei

Individual Drugs	Fluorescence (Color/Maxima), nm
Sesquiterpene lactones	
Artemisinine	Blue/440
Azulene	Blue/425
Gaillardine	Blue/420
Tauremizine	Blue/440
Anthocyanins	
Pelargonidin	Green/520
Alkaloids	
Atropine	Blue/415
Berberine	Yellow/540
Casuarine	Blue/475
Chelerythrine	Green-yellow/510, 540
Colchicine	Blue/440
Glaucine	Blue/440
Physostigmine	Blue/410
Rutacridone	Orange/595
Sanguinarine	Orange/595
d-Tubocurarine	Blue/415
Yohimbine	Blue/455

Alkaloids. Some quaternary benzo[c]phenanthridine alkaloids (sanguinarine and chelerythrine, etc.) may fluoresce within cells, binding to DNA (Slaninova et al. 2008). They show the nucleus architecture similarly to common DNA dyes. When they are excited by argon lasers at 488 nm, a scientist can see fluorescence in the 576–700 nm range. The alkaloids can be used as fluorescent DNA probes for living cells in both fluorescence microscopy and flow cytometry, including multiparameter analysis.

The yellow alkaloid rutacridone is the active chemical of rue roots, which may participate in root–root interspecies interactions (Alliotta and Cafiero 1999) and inhibit the development of seeds. Treatment with rutacridone led to binding with the nucleus of *Hippeastrum hybridum* (Roshchina 2002). After histochemical staining, the nucleus and chloroplasts fluoresce in green, unlike the cytoplasm, which fluoresces orange or yellow. The analysis of rutacridone interaction with acceptor cells showed the shift of orange fluorescence with a maximum at 580–586 nm in orange to green emission in the region of the nucleus in pollen of *H. hybridum* as well for the nuclei and chloroplasts of vegetative microspores of *Equisetum arvense* (Roshchina et al. 2011). Similar results were obtained when this alkaloid used to stain isolated nuclei from white petals of *Philadelphus grandiflorus* and *Matricaria recutita*. The nuclei fluoresced green (Figure 4.6) and had two fluorescence spectrum maxima at 460–480 nm and 585–590 nm, whereas only one maximum (585–590 nm) was observed in the cytoplasm of cells or in the rutacridone solution. The staining of isolated chloroplasts with rutacridone decreased their red fluorescence (due to chlorophyll) and enhanced the green emission, which supports the hypothesis that such chemicals can enter chloroplasts, possibly at DNA binding sites (Roshchina et al. 2011).

Anthocyanins. In a paper by Roshchina and coworkers (2011), the fluorescence spectrum (excitation by laser 405 nm) of *Hippeastrum hybridum* pollen stained with the anthocyanin pelargonidin had maximum 450 nm and shoulder 520 nm. The common view of the object fluoresced in green, and mainly the nucleus emitted in the center (Figure 4.6b). Thus, pelargonidin was able to penetrate into the cell and color the nucleus.

Azulenes. Synthetic azulene and chamazulene from *Matricaria reticuta* were tested on single plant cells (Roshchina 2004, 2008, 2011). Staining vegetative microspores of horsetail *Equisetum arvense* (chloroplast-containing models) with azulene induced blue (excitation360–380 nm) or blue-green (excitation 420–430 nm) fluorescence of nucleus and chloroplasts (i.e., DNA-containing structures) instead of the red fluorescence seen in the control. The blue fluorescence intensity was also seen in the fluorescence spectra of pollen nuclei of *H. hybridum* and *Vallota speciosa*. Also, treating isolated chloroplasts with azulene increased the blue fluorescence intensity and decreased the red emission of organelles. The possible target of azulene is DNA-containing structures, because the compound interacts with nuclei and chloroplasts. Fluorescence in the presence of azulene has been shown earlier in nuclei isolated from white petals of flowers of *Philadelphus grandiflorus* and *M. recutita* stained with azulene (Roshchina 2005a). The possible target of azulene is DNA-containing structures, because the compound interacts with nuclei (Roshchina et al. 2011). The fluorescence of cell nuclei in the presence of azulene was shown earlier (Roshchina 2004, 2008, 2011a). The extract from *Artemisia absinthium*, rich in chamazulene also induced blue fluorescence of nuclei (Roshchina et al. 2011).

Modeling of dye interactions with components of nucleic acids: The interaction of fluorescent drugs with the components of nucleic acids—nucleosides, that is, pyrimidine and purine bases bound to ribose or deoxyribose, thymidine, cytosine, adenosine, and guanosine as well as ribose(deoxyribose) itself—was studied using films of compounds on glass slides (Roshchina et al. 2011). Using films to model solid systems fluorescing in the 380–410 nm region was due to several circumstances: (a) in nature, nucleic acids are in solid form and (b) in the solutions of the nucleosides and sugars, there was very little fluorescence, if any. In the fluorescence spectra of films from 2 mg, every nucleic acid component was treated with 10^{-5} M dyes. The emission excitation was 360 nm. At short time exposure (<1 h), the blue anthocyanin fluorescence was quenched only in treatment with ribose (deoxyribose). This correlates with data for whole cell staining, when the nucleus transformed fluorescent pelargonidin into non-fluorescent cyanidin. The binding site of cyanidin is proposed to be the phosphorus between two deoxyribose residues of the DNA near the contact of ribose (deoxyribose) with nucleotides (Jaldappagari et al. 2008). The anthocyanin–DNA copigmentation complex becomes the protection against oxidative damage (Sharma and Sharma 1999). Unlike results with pelargonidin, among the films stained by azulene, only guanosine increased the fluorescence at 420–430 nm (Roshchina et al. 2011), while other nucleosides did not interact. The DNA–azulene complex emits in blue-green with 440 and 520 nm maxima (Roshchina 2004). Guanosine has been considered as a potential target for the chemical. Another picture was after staining by rutacridone. In this treatment, blue-green emission was observed mainly for cytosine (maxima at 415 and 445 nm in the fluorescence spectra) and to a small degree for ribose (a smooth peak at 445 nm was seen), while the solution of rutacridone had no maxima in the spectral region. Green (or blue-green) fluorescence of nuclei and pure DNA has also observed in the presence of rutacridone. Cytosine appears to be one of the targets for rutacridone on the nucleic acid. The tested compounds were considered as histochemical dyes and used for the indication of DNA-containing cellular structures in various cellular compartments. The alkaloid rutacridone and sesquiterpene lactones (gaillardine and grosshemine), which inhibited the pollen germination, passed through the plasmalemma into the cell, and the nucleus became fluorescent in green (rutacridone) or blue (the above-mentioned sesquiterpene lactones). The substances also reacted with DNA or nucleic acid–protein complexes, inducing similar fluorescence. After similar histochemical staining, the nucleus and chloroplasts fluoresce in green, unlike the cytoplasm, which fluoresces in orange (Figure 4.6).

In comparison with rutacridone, other alkaloids, which have anticholinesterase activity, such as physostigmine, berberine, and sanguinarine, concentrated and fluoresced with orange color on the surface of the cell. DNA was the potential target for these chemicals, in particular as sugar ribose.

4.5 REACTIVE OXYGEN SPECIES (ROS) IN PLANT TISSUES *IN VIVO*

An increase in the amount of reactive oxygen species (ROS), such as free radicals and peroxides, in cells and intact tissues is an indicator of stress. Fluorescence allows simpler and more selective *in vivo* ROS detection (Ortega-Villasante et al. 2018). Notwithstanding the strong autofluorescence of many secondary metabolites (see Chapter 2), it may take place in combination with appropriate fluorescent probes, subcellular monitoring, and quantification of ROS dynamics. In green tissues, chlorophyll is the major contributor to autofluorescence, but significant interference also comes from cell wall components (e.g., cellulose and lignin) and other colored molecules and pigments (e.g., carotenes, xanthophylls, flavonoids, anthocyanins, alkaloids, etc.). Although fluorescent methods are not widely used in pharmacy, here one should note the future potential for the determination of superoxide radical and hydrogen peroxide.

The use of fluorescent probes leading to ROS detection by dye-based probes proceeds as follows. In the presence of superoxide (superoxide radical), a similar probe, DHE or dihydroethidium (2,7-diamino-10-ethyl-9-phenyl-9,10-dihydrophenanthridine) transformed into a compound with fluorescent phenanthrene-like rings. The dye permeates cell membranes, producing the red fluorescent 2-hydroxyethidium (2-OH-E+), which on excitation at 500 nm, leads to the appearance of an emission peak at 600 nm. Alternatively, DHE can be catalytically oxidized to E+, which features very similar fluorescence characteristics. The DHE probe is mainly applied to cultured plant cells.

Hydrogen peroxide detection may be carried out with 2,7-dichlorodihydrofluorescein (DCF)-based probes or dihydrorhodamine 123 (DHR)-based dyes. Probes based on DCFH-DA (dichlorodihydrofluorescein diacetate) cross the plasma membrane, and esterases cleave the acetate groups to produce DCFH, which is oxidized to DCF, an anthracene-like chemical structure that is capable of fluorescence. The same scheme can be applied to DHR-based probes; at the top of DCF molecules formulae are amino ($-NH_2$) groups in the case of rhodamine 123. The third possibility is to use Amplex Red® dye, which undergoes oxidative deacetylation in the presence of cellular peroxidases, leading to the production of fluorescent resorufin.

CONCLUSION

The application of fluorescent methods in histochemistry and cytochemistry enables the determination of certain medicinal compounds in secretory structures containing drugs. In this case, besides phenols and terpenes (predominantly seen with color reactions), a scientist may see biogenic amines assayed after staining with reagents that induce fluorescence of these neuromediators/neurotransmitters. It is possible to avoid lengthy biochemical procedures by using a small amount of pharmaceutical material in microscopic analysis. The viability of fresh cells can also be estimated based on the state of nuclei stained by various fluorescent dyes. In the future, industry may prepare and sell new fluorescent probes for drugs, including for the determination of reactive oxygen species (ROS).

5 Fluorescence in Phytopreparations

The main information dealing with plant excretions has been given in an earlier monograph (Roshchina 2008), where intact secretory cells were analyzed and compared with the released secretions. The natural secretions released include slime, nectar, resin, and oils seen on the plant surface or evacuated on the glass slide of the microscope. In this chapter, we consider the medicinal aspects of plant secretions applied to pharmacy. Most known approaches for phytopreparations deal with the various separation procedures for natural products (Cseke et al. 2006). Fluorescence may be useful in the express analysis of ready phytopreparations such as tinctures, oils, and juices. Researchers can estimate the freshness of the pharmaceutical material. The first steps in fluorescence analysis for fresh pharmaceutical materials were on *Citrus* peel (Momin et al. 2012). In the publication, the main maxima in the emission spectra of different *Citrus* varieties were from 500 to 540 nm when excited by light 360–380 nm. The fluorescent method is also used for the study of leaf salt glands and salt secretions of *Limonium bicolor* and *Aegialitis rotundifolia* as well as trichomes and stomata of *Arabidopsis* under excitation by UV light 330–380 nm (Yuan et al. 2013). The amount and density of the glands on the leaves have also been determined using this method.

Interactions between biological molecules enable life (Luzarowski and Skirycz 2019). This is true both for natural secretory structures and for extracts of medicinal plants. As well as color and smell, the fluorescence of flowers may attract bees (Gandia-Herrero et al 2005; Iriel and Lagorio 2010a, b). Flavonoids play a major role in the process, but some pigments, such as betaxanthins or betacyanins, also can contribute to the emission in blue (maximum 475–480 nm) and green (510–520 nm). These compounds are peculiar to *Bougainvillea*, *Celosia*, *Gomphrena*, and *Portulaca*, whose extracts are used in folk medicine of tropical countries.

The significance of a cell-wide understanding of molecular complexes is thus obvious, although only protein–metabolite (small molecules) relationships are studied. A similar approach enables detection of the differences between extracts, admixtures (including artificial additions that differ from drugs known for intact secretory cells *in situ*), and counterfeits, if the probes fluoresce.

It must be kept in mind that the fluorescence of phytopreparations may depend on the solvents. As for natural media where the drug is present, modeling shows that within secretory cells, various components, both hydrophilic and hydrophobic, occur and can interact with each other (Roshchina 2008). The natural solvents for the main fluorescent metabolites in mixtures are water, oils, and alcohols in different secretory structures of medicinal plants. To a certain degree, we are able to model the situation when we see how isolated individual compounds fluoresce in different spectral regions depending on the presence of other components of natural secretions. A spectral parameter may apply for testing the quality of the phytopreparations.

5.1 PREPARATIONS FROM INDIVIDUAL PLANT COMPOUNDS AND MIXTURES

In many cases, internal secretions are rarely excreted from secretory structures, and special methods of analysis are needed. Liquid components of various secretions may be extracted by the solvents and compared with the fluorescent components in intact secretory cells *in situ*. The duration of extraction is an informative indicator of where the secretions are located: in external structures on the tissue surface or within the tissue. Unlike the above-mentioned natural excretions, water extracts may additionally include water-soluble cellular components, which are not contained in secretory cells. In other words, the fluorescence of these extracts may serve only as a model system for some secretions. Examples are as follows: (1) washings or leachates from the surface (leaching for no more than 1 hour) or infusions of the tissue over a longer time; (2) extracts by organic solvents, both those encountered in plants and artificial ones. The fluorescent properties of many tinctures and other phytopreparations are of interest, because the main emitting components also may be known as active matter.

5.1.1 WATER EXTRACTS AND TINCTURES

Some secretions may be evacuated by water. The extracts (washings or leachates from the surface, leaching for no more than 1 h, or infusions from the tissue over a longer time) serve as a model system. Washings (leachates) and infusions from flowers and leaves of the various species contain flavonoids, which fluoresce with maximum 450–470 nm, specific to the secretory cells of the leaf, which fluoresce brightly in the blue spectral region (Roshchina 2007 b, 2008). In some cases, the fluorescence appears to differ from the blue-emitting glands. Water extracts from bud scales of *Populus balsamifera*, *Betula verrucosa*, and *Aesculus hippocastanum* contain phenolic components fluorescing in blue at 450–470 nm (Roshchina and Melnikova 1995). Some examples of fluorescence from water extracts of herb plants' leaves are represented in a monograph (Roshchina 2008). The washings or leachates from the leaf epidermal tissues with secretory cells show fluorescence with one maximum at 450–460 nm for three plant species—*Urtica dioica*, *Symphytum officinale, and Hypericum perforatum*—although there is a difference in the chemical composition. This may be due to the presence of phenolic compounds, which fluoresce in blue at 450–460 nm in stinging emergences of *Urtica dioica* and in slime of *Symphytum officinale* as well as in oil cavities and ducts of *Hypericum perforatum* (Roshchina et al. 1997a, b, 1998a; Roshchina 2001). Comparing the fluorescence of intact pollen and water extracts from the sample enabled us to evaluate the contribution to luminescence of various phenolic components too. In particular, anthocyanins can emit not only in blue with a maximum at 450–470 nm, but also in orange-red with maximum 585–605 nm on excitation by light at 360–380 nm (Roshchina and Melnikova 1996; Melnikova and Roshchina 1997).

Special investigations have been devoted to the connections between the content of fluorescent components of extracts and their medicinal applications. Flavonoids play a major role in the process, but some pigments, such as betaxanthins or betacyanins, can also contribute to the emission in blue (maximum 475–480 nm) and green (510–520 nm). These compounds are also specific to *Portulaca*, whose water and ethanolic extracts are used in folk medicine in tropical countries (Gandia-Herrero et al. 2005a, b; Iriel and Lagorio 2010a, b). *Portulaca oleracea*, as an edible and medicinal plant, is of considerable importance to the food industry and also possesses a wide spectrum of pharmacological properties, such as neuroprotective, antimicrobial, antidiabetic, antioxidant, anti-inflammatory, antiulcerogenic, and anticancer activities, which are associated with its diverse chemical constituents, including flavonoids, alkaloids, polysaccharides, fatty acids, terpenoids, sterols, proteins, vitamins, and minerals. The aqueous extract of *P. oleracea* also prevents diabetic vascular inflammation, hyperglycemia, and diabetic endothelial dysfunction in type 2 diabetic db/db mice, suggesting its protective role against diabetes and related vascular complications (Zhou et al. 2015). The main pharmacological properties related to the neuroprotective activity of the plant are that it can scavenge free radicals and antagonize rotenone-induced apoptosis of neurons, dopamine depletion, and complex-I inhibition in the striatum of rats. There is a supposition that *P. oleracea* may be a potential neuroprotective candidate against Parkinson's disease. The extract protects nerve tissue/cells from hypoxic damage, probably by elevation of glycolysis and hypoxia inducible factor-1 expression levels. An increase in neuron viability and the induction of mRNA and protein expression of endogenous erythropoietin have also been reported. β-Cyanin evidently inhibits D-galactose-induced neurotoxicity in mice, which at the doses of 50 and 100 mg/kg, upregulates the activities of superoxide dismutases, catalase, glutathione reductase, and glutathione peroxidase. Extract of *P. oleracea* significantly inhibited acetylcholinesterase (AChE) activity at a final concentration of 100 µg/mL with the IC_{50} value of 29.4 µg/mL. The use of the extract, containing AChE inhibitors, is a promising treatment strategy for Alzheimer's disease. Furthermore, it possesses antimicrobial effects too.

5.1.2 JUICES

As well as washings and infusions, juices are recommended as natural drug products. One example (Figure 5.1) is the juice of *Aloe* leaves (family Aloaceae), first used as a biogenic stimulator in ophthalmology treatment by Professor Filatov. He wrote about it in the first monograph (Filatov 1955) where it was determined that tissue stimulators from both animals and plants as sterile homogeneous suspensions or extracts of the tissue contained compounds able to stimulate natural living processes in organisms. Similar preparations from the cornea, lens, spleen, and placenta were applied widely in the USSR in medicine and livestock raising. Moreover, animal tissue preparations made according to the Filatov method and stored in the cold may regulate plant growth and development (Blagoveshchenskii 1956; Roshchina V.D. 1964) as well as the biological activity of yeast (Roshchina V.D. 1962). Today, extracts from *Aloe arborescens* and other *Aloe* species are widely used in clinical trials (Singab, Abdel-Naser et al 2015; Radha, Maharjan H.; Laxmipriya, Nampoothiri 2015). For

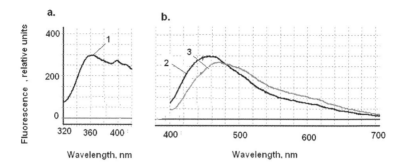

FIGURE 5.1 The fluorescence spectra of juice from *Aloe arborescens* leaves. (1) excitation spectrum for the emission 460 nm, (2) and (3) emission spectra on excitation at 360 and 380 nm, respectively.

example, they have been used in the treatment of wound healing (Shedoeva et al. 2019; Singh and Sharma 2020) or cancer (Lissoni et al. 2009).

As seen in Figure 5.1, the excitation spectrum of the *Aloe* preparation includes a main maximum at 360 and a shoulder at 380 nm, while the maxima of fluorescence on excitation in the spectral region were at 460 and 470 nm, relatively, which are due to flavonoids. However, the fluorescence spectra are very prolonged, and a small shoulder was seen in the orange-red region, linked with the presence of anthraquinones such as Aloe-emodin (Golovkin et al. 2001).

5.1.3 Preparations by Extraction with Organic Solvents

Secretory components related to water-insoluble substances can be extracted with more hydrophilic or more lipophilic organic solvents. For tinctures, non-hazardous solvents are used, in particular, ethanol (often 70–75% in mixture with water), which can dissolve most hydrophilic and hydrophobic compounds. The content of components is wider than in water tinctures (see Section 5.1.1 for an example in *Portulaca oleracea*). This ethanol extract decreases the activity of caspase-3 in neurons while reducing serum levels of neuron specific enolase in hypoxic mice. Pathological damage caused by hypoxia is due to cyanidin flavonoid and betalanins, which fluoresce in blue-green or even yellow (Gandia-Herrero et al. 2005; Iriel and Lagorio 2010).

The bud scales of woody plants such as *Populus balsamifera*, *Betula verrucosa*, and *Aesculus hippocastanum* are rich in lipophilic compounds (Roshchina V.D. and Roshchina V.V. 1989; Roshchina V.V. and Roshchina V.D. 1993). In particular, their ethanol and acetone extracts fluoresce in blue at 450–470 nm (Roshchina and Melnikova 1995). The ethanol extracts of lipophilic pollen compounds of the same species contained flavonoids and azulenes, which can fluoresce in the blue (maxima at 430–470 nm) spectral region, whereas carotenoids emit in green-yellow with maxima at 520–560 nm (Roshchina et al. 1995, 1997a, b, 1998a).

Many lipophilic products of secretions are in oil ducts, resin ducts, and other terpenoid-containing secretory structures. Primarily, they may be extracted from the structures by non-polar solvents such as chloroform, petroleum ester, or others. (Then, the dried extracts as drugs for humans are mixed with natural oils or the innocuous medicinal solvent dimethylsulfoxide.) The extraction from the secretory cells, which are located on the surface, depends on the time of exposure to these solvents and the rate of solvent penetration into deeper tissues. At exposures lasting longer than 10 min, the extract may include more chlorophyll than is contained in secretory structures located on the surface. At exposures of 60 min, chloroform extracts the pigment from mesophyll after extraction for 10 min (Roshchina 2008). Extractions for 10 min from microsporangia with microspores (pollen) of *Pinus sylvestris* also contain blue-fluorescing compounds with maxima at 450–455 nm, while the extractions from *Pelargonium* leaves (enriched in oil glands) show a maximum 460 nm (10 min extract) and maxima at 410, 435, 450 nm (60 min extract) with the additional peak of chlorophyll. The extracts of leaves from *Artemisia absinthium* demonstrated weak maxima 430 and 460 nm after 10 min of exposure to the solvent, whereas after 60 min, the maxima were 414 and 430 nm, with shoulder at 465 nm. When the excitation was at 420–430 nm, all maxima, except that of chlorophyll, shifted to the longer-wavelength region of 460–490 nm. Thin-layer chromatography on silica gel or Whatman paper 1 separated a green band emitting in red (summed chlorophylls and azulenes) and a weak yellowish band (terpenoids), which fluoresced in blue with maxima at 430–435 and 460–465 nm.

Internal secretory reservoirs storing oils and resin are widespread in the Asteraceae family, and one example is demonstrated for *Solidago virgaurea* L. and *Solidago canadensis* L. (Lersten and Curtis 1989), which is used in medicine (Sonova 2015; Suleymanova et al. 2017). Intact oil reservoirs fluoresce in blue (maxima at 410 and 428 nm) when excited by ultraviolet (UV) at 360–380 nm or in blue-green (maxima at 450 and 470 nm) when excited by violet light (420 nm), as seen in the monograph (Roshchina 2008). Extraction for 10 min shows similar emission, but after longer extraction, some components with emission maxima at 485 and 528 nm appeared. The chlorophyll maximum at 680 nm increased more than 100 times, which was difficult to include in the same spectrum of the extract. Therefore, practically, 10 min extraction is desirable for releasing the oil secretory products from epidermal secretory structures without chlorophyll from mesophyll cells. A blue spot after thin-layer chromatography on silica gel (perhaps azulenes) fluoresces in blue with maximum 450 nm (Roshchina 2008). Azulenes are often encountered mixed with oils in the oil-containing reservoirs (Roshchina and Roshchina 2003, 2012).

Fluorescent phytopreparation usually characterizes the prevailing component, whose emission also reflects active matter. Extraction by organic solvents, mainly by ethanol, is applied for the preparation of tinctures. Figure 5.2 may serve as an illustration, where the fluorescence spectra of widely used medicinal preparations are considered. Here are the examples for *Calendula officinalis*, *Kalanchoe pinnata*, and *Crataegus oxyacantha* as well as the preparation Papilate (an oil mixture of 50 natural components). The fluorescence is excited by light at 360 or 380 nm. Under the first excitation, small maxima in *Calendula* extracts from flowers were seen at 417, 444, 465, and 515 nm, higher peaks at 493, 540, 565, 635, and 675 nm, and a shoulder at 595 nm. At the second excitation, the only clearly visible maxima were at 520, 570, 630, and 680 nm. Therefore, preliminary fluorescent analysis shows the presence of sesquiterpene lactones emitting in blue, a high content of carotenoids fluorescing in orange, flavonoids, and, excluding the chlorophyll maximum at 675–680

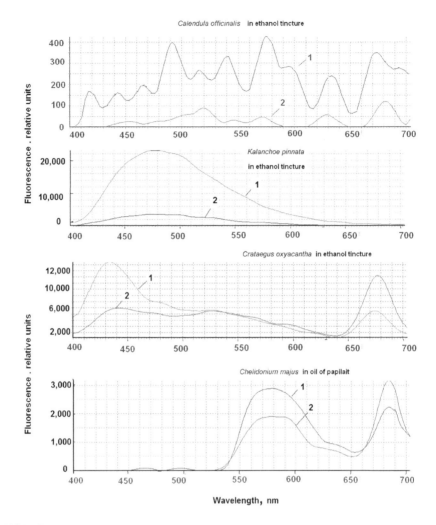

FIGURE 5.2 The fluorescence spectra of ethanol tinctures from flowers of *Calendula officinalis* (100 g flowers + 75% ethanol up to 1000 mL), leaves and shoots of *Kalanchoe pinnata* (760 mL of juice + 95% ethanol up to 1000 mL), fruits of *Crataegus oxyacantha* L. (100 g/L of 70% ethanol), and oil preparation Papilite (Papilait). (1) and (2) Excitation at 360 and 380 nm, respectively.

nm, unknown components emitting in red at 620–630 and 590 nm. *Calendula* tinctures are used in medicine. In the orange-colored flowers of this herb, there are many carotenoids (the emission maxima at 565, 570, and 590 nm), among them carotene, rubixanthine, lycopene, tsigroksanthin, violaxanthin, and others. As we have seen, flavonoids also are present, due to peaks at 465 and 490 nm. The medical purposes, including antiseptic, anti-inflammatory, and wound healing, require *Calendula* tincture preparation to be characterized. Besides, the extracts demonstrate vitamin and phytoncide properties.

In Russian (Golovkin et al. 2001) and Indian (Shruti et al. 2018) pharmacy, *Kalanchoe pinnata* (Lam.) Pers, a succulent plant originating from South Africa, is widely used in medicinal practice today. It has potential anticancer and insecticidal compounds—bufadienolides—in leaves and stems (Joseph et al. 2011), but is rarely used in World medicine yet (Morsy 2017). This perennial medicinal herb is popularly used in Brazil and other parts of the world to treat various inflammatory diseases (Almeida et al. 2006). Almeida with coworkers (2006) reported biological activities such as immunosuppressive effects, hepatoprotective activity, and acetylcholinesterase inhibition, as well as important protection against progressive infection with *Leishmania amazonensis* due to a minor vinylic aliphatic alcohol diglycoside, whose structure was proposed as the known 1-octen-3-*O*-a-L-arabi nopyranosyl-(1,6)-b-glucopyranoside I. This herb is also widely used against rheumatism and inflammatory processes and showed protection against the local effects (edema, hemorrhage, and necrosis). In the juice, cardiac glycosides, bryophyllin, and its derivatives have been found. Bufadienolides are a group of polyhydroxy C-24 steroids and their glycosides, containing a six-membered lactone (α-pyrone) ring at the C-17β position. From the pharmacological point of view, bufadienolides might be a promising group of steroid hormones with cardioactive properties and anticancer activity (Kolodziejczyk-Czepas and Stochmal 2017). Fluorescence spectroscopy is 10–100 times more sensitive than absorption photometry, but as yet it is only used for digoxin determination; excitation is at 340 nm, and the emitted fluorescence is measured at 420 nm (Morsy 2017). Also, other phytochemicals have effects against urinary tract, parasitic, and bacterial infections and are antiulcer actives and antidepressants. The anticancer effects are due to the presence of bersaldegenin-1,3,5-orthoacetate, which inhibits the growth of several cancer cell lines (Supratman et al. 2001). *In vitro*, it shows antibacterial activity toward *Staphylococcus*, *E.coli*, *Shigella*, *Bacillus*, and *Pseudomonas* bacteria. As seen in Figure 5.2, in the fluorescence spectra of the ethanol tincture of *K. pinnata* leaves, there is only single maximum 480 nm (highest at excitation by light 360 nm), perhaps due to flavonoids such as quercetin, kaempferol, and others that are widespread in the genus *Kalanchoe* (Golovkin et al. 2001). Juice from fresh shoots and leaves of *Kalanchoe*, in conjunction with other medical uses (physiotherapy and antibiotics), is also known as an effective healing and anti-inflammatory agent, widely used in medical practice in surgery, dentistry, gynecology, and ophthalmology and to treat various forms of keratitis. Until 1960, *Kalanchoe* as a herb was practically unknown in Europe. The history of its use in medical practice began the 1970s thanks to Ukrainian scientists (Kutsik and Zuzuk 2004). Then, in different scientific centers of the former Soviet Union, studies were carried out on the botanical characteristics, chemical composition, and pharmacological properties of biologically active compounds of plant species of the genus. Fluorimetry of cardiac glycosides in the species was not carried out, although as early as 1975, Avissar with coworkers presented an indirect method that used trichloracetic acid as a reagent with the compounds emitting in blue. In our preparation from *Kalanchoe*, there was only one maximum 475 nm specific to many phenolic compounds, especially the flavonoids patuletin, quercetin, and kaempferol, found in the species of the genus (Golovkin et al. 2001).

In the fluorescence spectrum of ethanol tinctures from fruits of the northern European hawthorn, *Crataegus oxyacantha* (Figure 5.2), the reader can see three maxima 435, 530, and 675 nm and shoulder 480 nm, which relate to sesquiterpene lactones, carotenoids, chlorophyll, and flavonoids, respectively (Roshchina 2008). This cardioprotective herb is used as a spasmolytic and cardiotonic in cardiology due to its active matter content of flavonoids and triterpenoid acids (Murav'ova et al. 2007). This confirms the contribution of many flavonoids to the visible fluorescence of the tincture.

The complex phytopreparation Papilite is based on the oil from *Chelidonium majus* that includes more than 40 other medicinal valuable components, such as propolis (a waxy glue made by bees), *Melaleuca alternifolia*, mimeo or shilajt, brakshun, oleum *Hippophae*, castorium of beaver, etc. This preparation has been applied to papillomas with visible success. Effective application occurs due to alkaloids such as chelerythrine, sanguinarine, and others. In Figure 5.2, besides maximum of chlorophyll peak 675–680 nm, one can see in the fluorescence spectra the characteristic maximum in the 540–600 nm range, which includes the contribution mainly of the above-mentioned alkaloids. In any case, the emission with a peak 575 nm of the main active drug alkaloid, chelerythrine, of *Chelidonium majus* (Golovkin et al. 2001), is predominant in the probe and show antifungal features (Parvu et al. 2013).

5.1.4 Medicinal Oils

Unlike the usual medicinal tinctures, natural oils contain and may be extracted only with hydrophobic compounds needed for pharmacy and cosmetics (Hüsnü et al. 2009). The medicinal properties of essential oils are known as being more widely applied for preparations of various ointments as well as in natural forms (Bassole' and Juliani 2012; Raut and Karuppayil 2014).They are currently of great importance to pharmaceutical companies, cosmetics producers, and manufacturers of veterinary products. Essential oils are found in perfumes, creams, bath products, and household cleaning substances. They are used for flavoring food and drinks. Some essential oils are known to act on the respiratory apparatus, which may be hazardous. On the whole, essential oils demonstrate antimicrobial activity (Pauli and Schilche 2009). This is especially valuable because essential oils may enhance the effect of biopolymers in wound healing (Pérez-Recalde et al. 2018).

The luminescent properties of essential oils have not been extensively investigated until now, when they have been analysed (Boschi et al. 2011). The increasing interest in optical imaging techniques and the development of related technologies have made possible the investigation of the optical properties of several compounds, among them melissa and lavender, which proved to be the most luminescent compounds after sunshine exposure. Momin with coworkers (2012) tried to record the fluorescence spectra of *Citrus* peel oil in the UV-VIS region. On excitation at 350–380 nm, the samples fluoresced in yellow at 490–540 nm, showing mainly tangeretin (flavone) as well as possibly carotenoid emission.

Oil phytopreparations on glass slides, represented in Figure 5.3, fluoresce with various degrees of intensity as seen under a confocal microscope and by a spectrofluorimeter. Maxima are shown in blue, such as 465 nm for lavender, *Lavandula*, 470

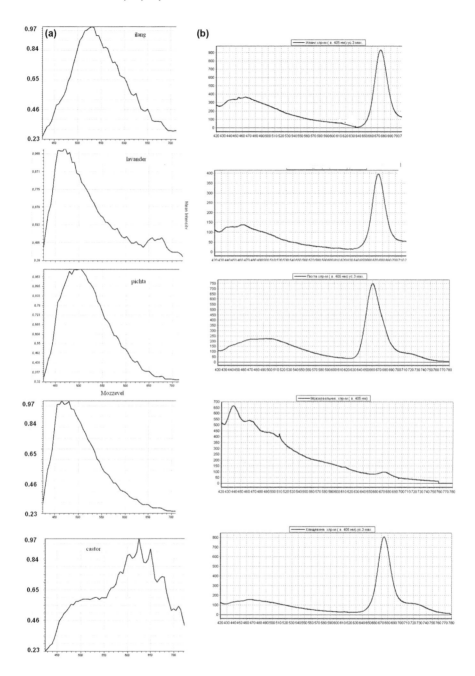

FIGURE 5.3 The fluorescence spectra of medicinal and cosmetic oils. (a) On slides by confocal microscopy, excitation by 405 nm laser; (b) extract from oils by chloroform (1:10 v/v), spectrofluorimetry, 0.2 cm cuvette, excitation at 405 nm.

nm for juniper, *Juniperus* (the spectrofluorimeter also shows 438 nm), and 460 nm for ylang, *Cananga* (the spectrofluorimeter also shows 438 nm). In blue-green, maxima at 490–525 nm are clearly seen for oils of ylang, *Cananga odorata*, fir, A*bies balsamea*, and castor bean, *Ricinus*. In most samples of Figure 5.3b, there is maximum 675 nm, peculiar to chlorophyll, which is better seen by the spectrofluorimeter, perhaps due to a deep layer of oil in the cuvette. As well as chlorophyll, essential oil of castor has peaks at 590 and 620 nm in the fluorescence spectra recorded by confocal microscopy.

Ylang-ylang (*Cananga odorata* Hook Fil.et Thomson) belongs to the family Annonaceae (maybe the family Lamiaceae) and is grown in its natural habitat in Burma, the Philippines, and Indonesia (Borneo, Java) and cultivated in plantations. The fragrance of essential oil of ylang-ylang calms down the nervous system, resulting in complete relaxation, relieves nervousness, and is a good help against children's fears. It is also an active aphrodisiac, able to create a romantic atmosphere and reinforce positive emotions. The flower oil of ylang-ylang contains many and various mono- and sesquiterpenes (Gaydou et al. 1986). Among them are α-cendrene, α-amorphene, γ-bisabolene, γ- and sigma-cadinols, p-methylanisole, α-murolene benzyl benzoate, geranyl acetate, and others—up to 50 components. As a whole, the terpenes weakly fluorescence in intact flower secretory cells, but natural oils prepared by distillation and their extracts by chloroform differ in the maxima in the fluorescence spectra (Figure 5.3). The oil on subject glasses or slides has a peak 540 nm, while the chloroform extract shows two peaks: 470 nm (terpenes) and 675–680 nm (chlorophyll as admixture).

Lavender essential oil, obtained by distillation from the flower spikes of certain species of the *Lavandula* genus in the Lamiaceae family. is also rich in various terpenes. It is used as a perfume, for aromatherapy, and for skin application in massage therapy for inducing relaxation (Elshafie and Camele 2017). Two forms are distinguished: lavender flower oil, a colorless oil, insoluble in water, having a density of 0.885 g/mL; and lavender spike oil, a distillate from the herb *Lavandula latifolia*, having a density of 0.905 g/mL. Like all essential oils, it is not a pure compound; it is a complex mixture of phytochemicals, including linalool and linalyl acetate. The essential oil extracted from *Lavandula officinalis* Chaix. showed strong antibacterial and antifungal properties. As seen in Figure 5.3, the natural oil and its chloroform extracts show a maximum in blue at 465–470 nm, specific to most of the terpenes and minor phenols contained in secretory cells (Roshchina 2008). However, the chloroform extract shows an additional maximum at 675 nm, due to chlorophyll, which is evidence of admixture.

Juniper (*Juniperus communis*) berry essential oil (JEO) is traditionally used for flavoring and medicinal purposes. It also known as the common juniper, is a dioecious aromatic evergreen shrub, and has been traditionally used in many countries as a diuretic, antiseptic, and digestive and as a flavor to aromatize certain alcoholic beverages (Falasca et al. 2016). A total of 90 components have been detected, and among them, remarkable qualitative and quantitative differences have been observed in the chemical components during the ripening stages, from the green unripe berries to the bluish-black berries harvested at full maturity. The monoterpene hydrocarbons decrease during ripening, with a progressive increase in sesquiterpenes such as germacrene D (12.29–17.59%) and β-caryophyllene (7.71–8.51%), which are the major components in ripe berry essential oils. This oil and its major active component alpha-pinene have been studied for antimicrobial, antifungal, antiproliferative, anti-inflammatory, and anticancer activities in a variety of settings (Bais et al. 2014). Admixture may include flavons (Bais et al. 2016). In addition, it has gained increasing popularity for skin health purposes. However, a literature search conducted by us showed no published studies regarding biological activity in human skin cells.

Also of great interest is castor bean, whose yellowish oil contains chlorophyll, as shown (van Gurp et al. 1989). We also confirm the presence of its contribution to a maximum at 675–680 nm both as the native oil and as a chloroform extract, but there are other maxima in the more short region. Usually, this emission could be linked to carotenoids (Roshchina 2008). Achaya with coworkers (1964) showed the composition of castor oil as follows (as wt %): palmitate 1.2, 0.9; stearate 0.7, 1.2; arachidate 0.3, 0.2; hexadecenoate 0.2, 0.2; oleate 3.2, 3.3; linoleate 3.4, 3.7; linolenate 0.2, 0.2; ricinoleate 89.4, 89.0; and dihydroxystearate 1.4, 1.3. Oxidative cleavage of purified methyl ricinoleate indicated that the double bond was exclusively in the 9–10 position. Castor oil glycerides were fractionated using 90% ethanol and commercial hexane as solvents (Achaya et al. 1964). Analysis of pooled fractions showed (as mole %) triricinolein 68.2, diricinoleins 28.0, mono-ricinoleins 2.9, and nonricinoleins 0.9. According to fluorimetry of edible oils in 2015, issued by PhysicsOpenLab, the fluorescence spectrum of castor oil excited by UV emission at 405 nm has several maxima: 410, 479, 524, 539, and 673 nm. The large maximum 479 nm presumably relates to the products of lipid peroxidation but alkaloids and flavonoids may contribute (Kang et al. 2004). Seed oils, in contrast, all show to varying degrees a clear fluorescence at about 470 nm, a sign of the content of peroxides resulting from oxidative degradation of fatty acids, both because they are supposedly hot-worked and because of the lower content of antioxidant molecules. These features make seed oils less suitable for use at high temperatures. In the spectrum of "Extra Virgin" olive oil, obtained by cold pressing, one can notice the absence of the products of peroxidation of fatty acids, which fluoresce at about 470 nm. This happens both because the oil is cold-worked and because the high content of natural antioxidants (carotenes and polyphenols) prevents the oil from oxidative degradation. The presence of the fluorescence peak of chlorophyll is clear.

On the whole, we have little information about the autofluorescence of medicinal oils. A small amount of data has been contributed by food science investigators, mainly for olive oil (Sikorska et al. 2012), which are used in medicinal preparations (unguents, ointments) as well. Some information deals with the UV spectral region, where at UV excitation (270 nm) of olive oil and its components, the emission ranged from 360 to 420–460 nm. If the sample included admixtures, such as phenolic compounds (gallic and other acids) or chlorophyll, their fluorescence, excited by light at 270 or 436 nm, also registered. Kongbonga with coworkers (2011) concluded that the main edible oils fluoresce in the visible spectral region when excited by light at 370 nm, and the maximum at 525 nm is due to vitamin E.

5.2   MODEL OF POSSIBLE INTERACTIONS OF METABOLITES IN SECRETORY CELLS AND PHYTOPREPARATIONS

Secretions in plant secretory cells are complex mixtures of different components. Here, they interact with each other, and the relationships result in various fluorescent characteristics. A similar picture may occur in phytopreparations after extraction of the compounds and formulation of the final drugs for a market. One approach to the analysis of their fluorescence may be to model the interactions using a mixture of individual compounds. In the following, we consider some examples of the interactions between water-soluble components or ethanol-soluble components used in phytopreparations. In the first case, there may be interactions between phenols such as anthocyanins and phenolic acids or between alkaloids and aromatic acids.

5.2.1 WATER SOLUTIONS

Figure 5.4 demonstrates how combinations of phenolic acids like aromatic acids and coumarins, or phenolic compounds and alkaloids, influence fluorescence spectra.

As seen in Figure 5.4, water solutions of anthocyanins as flavonoids show the fluorescence maximum at 425–430 nm, but the spectrum changes after the addition of phenolic acids such as o-coumaric acid. The spectra of the fluorescence in this case contain two maxima: 430 nm (the earlier peak specific to anthocyanins was 420 nm) and 515 nm (individual coumaric acid has peak 510 nm). Unlike the experiments with coumaric acid, in a mixture of anthocyanin with rutin (dissolved in basic solution), there are no maxima. Individual flavonol glycoside rutin fluoresces in yellow-orange with maximum 595–600 nm and anthocyanin may quench the emission in the mixture.

Unlike their combinations with coumarins, anthocyanins interacting with the alkaloid sanguinarine show no change in the fluorescence maximum of the compound. Similar behavior was seen with the alkaloid berberine, which is similar in chemical nature to sanguinarine.

Another picture was observed in the interactions of phenolic acids (Figure 5.4). In a mixture of ferulic acid and coumaric acid (maxima of the emission are 460 nm and 510 nm, respectively, for individual compounds), one can see an increase in height of maximum 460 nm. It is interesting that contacts of basic (pH 8.0) solution of flavonoid rutin with ferulic acids or the alkaloid berberine lead to a short-wave shift in the rutin maximum at 595 nm, while sanguinarine does not have this effect.

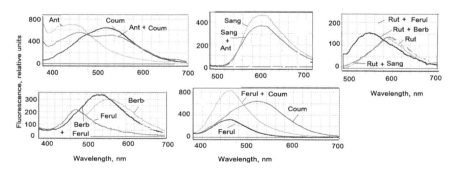

FIGURE 5.4   The fluorescence spectra for the interactions of water-soluble compounds (1 mg/mL). Ant: sum of anthocyanins from *Dianthus caryophyllus*, purified on Whatman 1 and dried in air; Berb: berberine; Coum: o-coumaric acid; Ferul: ferulic acid; Rut: rutin, pH 8.9; Sang: sanguinarine. Excitation: 360 nm for all substances, besides rutin and sanguinarine excited by light at 470 nm on subject glasses.

5.2.2 ETHANOLIC AND OIL SOLUTIONS

Interactions between compounds within model mixtures in ethanol and oil such as azulene-chlorophyll, anthocyanin-chlorophyll, or flavonoid anthocyanin–azulene— are shown in Figure 5.5. Azulene, soluble in ethanol, fluoresces from 380 to 430–460 nm depending on the excitation wavelength. The commercial azulene reagent also shows an additional maximum 675 nm, like chlorophyll. Unlike azulene, carotene is soluble only in a hydrophobic native medium such as oil. Films of azulene and carotene, air-dried on glass slides, fluoresce with maximum 430 nm in blue and 530 nm in greenish-yellow, respectively. In a mixture of both compounds in films, a peak 515 nm is observed. Anthocyanins fluoresce with maxima in blue (415 nm) and in yellow-red (595–600 and 650–660 nm). In mixtures of ethanolic solutions of azulene and anthocyanin, the spectrum does not contain a maximum at 675–695 nm. Of particular interest are the interactions between chlorophyll and azulene, the mixture of which shows red emission. This leads to a shift in the blue peak of chlorophyll, which moves to a shorter region of the spectrum, while the red maximum is at the same site as for chlorophyll. In other words, the changes are linked to the blue emission.

The changes related to the behavior of azulene in mixtures are not understood. It is difficult to explain the received spectral data. Perhaps, the non-stable maxima in blue and the presence of a peak in 675 nm are related to the structure of the compounds,

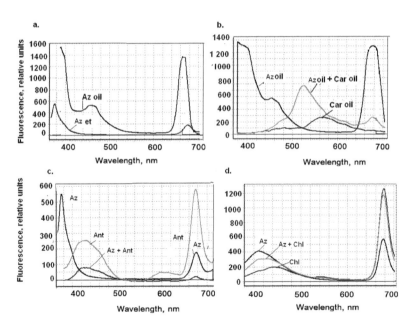

FIGURE 5.5 The fluorescence spectra for the interactions of water-insoluble compounds (1 mg/mL) for mixtures in a glass cuvette or in a film (a, c–d) on slides (a and b). Ant: sum of anthocyanins from *Begonia acetosa*, purified on Whatman 1 and dried in air; Az et and Az oil: azulene as a film in ethanol and in a mixture with mentholic oil; Car: carotene in sunflower oil; Chl: chlorophyll. Excitation: 360 nm for all substances.

with an 8-carbon chain and many double bonds. These fluorescence peculiarities must be kept in mind when formulating compositions of natural drugs. Azulenes are not products of distillations of essential oils; they are present in the chloroplasts of spring leaves in legumes (Roshchina 1999) and are found in pollen cells (Roshchina et al. 1995). According to Kasha's rule, fluorescence from organic compounds usually originates from the lowest vibrational level of the lowest excited singlet state (Si). Azulene, in exception to Kasha's rule, shows fluorescence from the state S_2 and displays the fluorescence originating from the second excited singlet, which is much stronger than that from Si (*Big Chemical Encyclopedia*). This behavior may be explained by considering that the azulene molecule has a relatively large S_2–Si gap, which is responsible for slowing down the normally rapid S_2 to Si internal conversion, such that the fluorescence of azulene is due to the $S_2 \rightarrow S_0$ transition. The fluorescence emission spectrum of azulene is an approximate mirror image of the S_0–S_2. Some fluorescence lifetimes are observed in picoseconds, although these are unusual cases. In organic molecules, the Sj–S_0 fluorescence has natural lifetimes of the order of nanoseconds, but the observed lifetimes can be much shorter if there is some competitive non-radiative deactivation (as seen for the case of cyanine dyes). A few organic molecules show fluorescence from an upper singlet state (e.g., azulene) and here, the emission lifetimes come within the picosecond time scale, because internal conversion to S and intersystem crossing compete with the radiative process. Fluorescence quenching may occur at the interaction of azulene with trans-stilbenes. Saltiel (1968) studied the effect of fluorescence quenching by azulene. Due to the structure of azulenes, which is rich in double bonds, they may interact with many compounds. For example, nanoparticles, like fullerenes, interact with azulene, and according to changes in the fluorescence registered by fluorescence spectroscopy, this has potential for pharmacy (Lorenzo et al. 2008).

Relations between the accumulation of azulenes and anthocyanins *in vivo* have been considered in only one publication (Bashirova et al. 2005). According to the data, there is a correlation between the increase of both compounds during the growth of medicinal plant species in the genus *Artemisia* belonging to various lines and sorts. This may be due to changes in metabolism, which depend on various factors. In *Achillea asiatica*, there is a correlation between anthocyanin and azulene synthesis. If the flowers were rich in anthocyanin, the accumulation of azulenes was 300% higher than control (Bashirova et al. 2003).

Similar changes may be seen in the fluorescence of secretory cells of the *Achillea* genus on excitation, for example, in *A. millefolium* (Roshchina et al. 2011b). Here were four examples of flower petals with different colors: white petal, white-rose petal enriched in various flavonoids, rose petal with weak anthocyanin pigmentation, and red-rose petal with bright anthocyanin pigmentation. The components were extracted by water, ethanol, and chloroform. Water extracts of all samples showed no emission, and neither did ethanol extracts of white petals. As petal color increased, their ethanol extracts demonstrated (1) weak emission in white-rose petal, (2) small clear maximum at 590–600 nm in weak rose petal, specific to some anthocyanins, and (3) high maximum at 590–600 nm in bright red-rose petal. When chloroform was used for extraction of more hydrophobic components, such as azulenes, in all samples, only one and the same maximum at 403–405 nm was seen. This maximum, perhaps, relates to proazulenes and azulenes that were both present in the species.

5.2.3 Correlation in Fluorescence between Cells and Extracts

Drugs in extracts and their localization in the raw phytomaterial may be compared for fluorescence between intact cells, cells after extraction by solvent, and the extract itself. Figures 5.6 and 5.7 and Table 5.1 represent examples of similar comparisons.

In Figure 5.6, fluorescent images of a leaf cell surface are seen before and after drug extraction for 1 h at excitation by light at 360–380 and 400–430 nm. The reader will observe how water-soluble and ethanol-soluble compounds have disappeared, liberating parts of the surface. In all cases, glandular structures, primarily seen in the control as dark spots, become more visible as secretory glands with a center and surrounding cells as a sheath. After water extraction, the whole surface weakly fluoresces, while treatment with ethanol opens internal cells containing chlorophyll

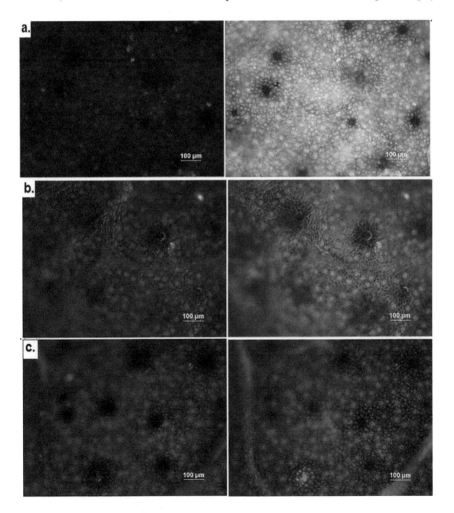

FIGURE 5.6 Fluorescent images of the surface of *Eucalyptus cinerea* before (a) and after extraction with water (b) and ethanol (c). Left: excitation by light at 360–380 nm, right: excitation by light at 400–430 nm.

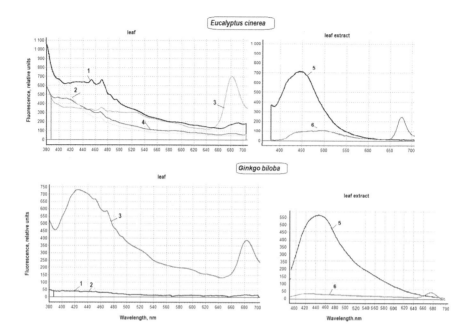

FIGURE 5.7 The fluorescence spectra of leaves and extracts from leaves of *Eucalyptus cinerea* and *Ginkgo biloba*. Water and ethanol extracts from the first species were at 2:100 and 10:100 w/v for 1 h, respectively, while the same extracts from the second species were at 3:100 and 5:100 w/v, respectively. For leaf: (1) control, (2) after extraction with water, (3) after extraction with ethanol, (4) after extraction with chloroform. For extracts: (5) water extract, (6) ethanolic extract.

(chlorenchyma, parenchyma), which emit in red. In officinal medicine, an ethanol tincture from various species of *Eucalyptus* is used. As seen in Figure 5.7, the control variant has emission maxima at 450, 470, and 685 nm. Water quenched the leaf emission such that only the smallest peaks at 450 and 470 nm were seen, while after the ethanol extraction, additional maxima at 530 and 680 nm were observed (flavins and chlorophyll, respectively). The spectrum of the water extract included a shoulder at 415 nm and the main maximum at 445 nm. In contrast, the ethanol extract emitted with peaks at 440, 580, and 680 nm. After treatment with the hydrophobic solvent chloroform, only maxima at 445 and 470 nm were seen in the leaf, as well as a chlorophyll peak at 680 nm (Table 5.1). After water extraction, leaf from *Ginkgo biloba* had no emission, but after the ethanol extraction, leaf fluoresced with maximum 423 nm, shoulder 470 nm, and maximum 685 nm. In the water extract, the emission demonstrated peak 445–450 nm, while ethanol extract – and small peak of chlorophyll 685 nm. The duration of extraction may result in a different picture of fluorescence, depending on the thickness of the leaf and surface components, which are extracted first.

Excitation by blue light at 360 and 380 nm induced mainly blue fluorescence of intact leaf tissue (Figure 5.7) with maxima at 430 nm (specific to terpenes such as proazulenes and azulenes) and 450–460 nm (due to some phenols and NADH/NADPH).

TABLE 5.1

Fluorescence Images and Fluorescence of Secretory Cells before and after Extraction with Various Solvents

Species	Organ	Fluorescence Maxima (nm) and Color		Solvent	Fluorescence Maxima (nm) and Color after Extraction	
		Ex. 360	Ex. 380		Ex. 360	Ex. 380
Actinidia chinensis	Leaf	455	460	Water	450	465
		Blue hairs			Weak blue hairs	
	Extract by water	460	460			
	Extract by ethanol	460 (small), 500, 680	460 (small), 500, 680	Ethanol		
Eucalyptus cinerea	Leaf	455, 470, 680	450, 470, 680			
		Greenish gland and dark gland (non-fluorescing)				
	Extracts by water	Shoulder 410, 450	460, shoulder 500	Water	Dark gland (non-fluorescing)	
	Extracts by ethanol	Shoulders 440 and 500, 680	Shoulder 460, 500, 680	Ethanol		
	Extracts by chloroform	450	460, shoulder 500	Chloroform	410, 470 (small)	470, 680 (small)
					Dark gland (non-fluorescing)	
Nerium oleander	Leaf	430, shoulder 455	430, 450, 465	Water	465	No emission
		Bluish hairs			Weak emission	
	Extract by water	460, shoulder 418	460			

In leaves of *Actinidia chinensis* and *Nerium oleander* under a luminescence microscope (Table 5.1), secretory hairs fluoresce due to surface components. After extraction with water, the hair emission decreased visually, and the fluorescence maxima shifted to shorter wavelengths. If to see the maxima of the extracts a reader marks that the 1-hour extraction completely evolve components fluoresced at 460 nm. It may be supposed that these are phenolic compounds, because they emit more intensely then NADH/NADPH usually do. When ethanol is used as a solvent, this maximum is also present, but new maxima at 500 and 680 nm appear due to flavins and chlorophyll, respectively. In the case of leaf of *Eucalyptus cinerea*, water extracts showed a maximum at 440–460 nm, possibly due to flavonoids. Nevertheless, ethanol extracts showed new maxima at 500 and 680 nm, like the two other-mentioned species. In the extract with chloroform, hydrophobic components extracted by chloroform had maxima in blue, and the glandular fluorescence disappeared completely. Only weak emission was seen with maxima at 410 and 680 nm. Terpenes, predominant in the leaf of the species, fluoresced with maxima < 460 nm, were extracted, and one can see their emission only in the chloroform extract.

5.3   SINGLE CELLS AS DRUG TRANSPORTERS

Today, an intensively developed conception relates to single cells as transporters of genes from one organism to another (Hwang et al. 2016). In this light, terrestrial plants have two to four times more ATP-binding cassette (ABC) transporter genes than other organisms, including their ancestral microalgae. Plants harboring mutations in these transporters exhibit dramatic phenotypes, many of which are related to developmental processes and functions necessary for life on dry land. ABC transporters multiplied during evolution and assumed novel functions that allowed plants to adapt to terrestrial environmental conditions.

This principle of transportation is specific not only to genes and proteins, but for transportation of small molecules through various membranes. This led us to the idea that it may be of interest for medicine. Earlier, there were some patents where authors tried to use the idea (Amer and Tawashi 1991, 1994; Atkin et al. 2005). Wolken with coworkers (2003) proposed spores as a potential material for pharmacy. Single living cells such as pollen, spores of spore-bearing plants, and even microalgae are natural containers for transport of drugs into the human body to enhance pharmaceutical delivery (Diego-Toboada et al. 2014). The authors noted that pollen grains and spore shells are natural microcapsules designed to protect the genetic material of the plant from external damage. The shell is made up of two layers: the inner layer (intine), made largely of cellulose, and the outer layer (exine), composed mainly of sporopollenin. The relative proportions of each vary according to the plant species. The structure of sporopollenin has not been fully characterized, but different studies suggest the presence of conjugated phenols, which provide antioxidant properties for the microcapsule and UV protection to the material inside it. These microcapsule shells have many advantageous properties, such as homogeneity in size, resilience to both alkalis and acids, and the ability to withstand temperatures up to 250 °C. These hollow microcapsules have the ability to encapsulate and release actives in a controlled manner. Their mucoadhesion to intestinal tissues may contribute to the extended contact of the sporopollenin with the intestinal mucosa, leading to an increased efficiency of delivery of nutraceuticals and drugs. The hollow microcapsules can be filled with a solution of the active or active in a liquid form by simply mixing both components together and in some cases, operating a vacuum. The active payload can be released in the human body depending on pressure on the microcapsule, solubility, and/or pH factors. Active release can be controlled by adding a coating to the shell or by co-encapsulation with the active inside the shell. Encapsulation and release of active compounds thus became of medicinal interest (Diego-Toboada et al. 2014).

Pollen particles, collected by bees, and spores, for example, as obtained from *Lycopodium clavatum* (club moss), are regarded as being beneficial to health and can be purchased in most health food shops. All the particles from a single species of plant have the same size and decorations. This morphology has been regarded as being so applicable to drug delivery that artificial pollen particles have been produced using silica (Guan et al. 2008; Cao and Li 2009), calcium carbonate or calcium phosphate (Hall et al. 2003), and iron oxide (Goodwin et al. 2013). However, these particles lack the elasticity of exines extracted from plant pollen and spores and are difficult to make free from defects. It is possible to attach drugs to the outside of complete particles, but the loading is restricted, the drug receives little protection, and its release is relatively uncontrolled. For over 10 years, some authors (Diego-Toboada et al. 2014) have developed a novel drug delivery system using the emptied shells of these particles and refilling them with an active ingredient. In this way, the hollow center gives a relatively high loading, and the shell itself provides protection. The use of coating material inside or outside the shell can provide controlled release. The microencapsulated active (Paul et al. 2009) can be used orally, by inhalation (very small particles only), topically, or in micro reactors (Paunov et al. 2007).The external shells of such materials have many properties, such as good elasticity, physical and chemical resistance, UV shielding capability, and antioxidant activity (Diego-Toboada et al. 2014).

The possibility of transport through the exine of microspores such as pollen or spores of spore-bearing species enables the structures to be used for drug microencapsulation (Diego-Taboada et al. 2013).

5.3.1 Pollen as Container That Transports Drugs Located in Single Cell

Pollen grains which are generative, microspores with male gametes are important for humans for many reasons. First, they define the normal breeding of forests and herbs, economic and medicinal plant species. Most significant is the maturing and fertility of pollen as the male gametophyte for the fertilization of the female gametophyte, leading to the formation of fruits and seeds (Stanley and Linskens 1974; Kabailiene 2010; Heslop-Harrison 2014). Pollen grains are carriers of the plant's genetic material needed for pollination, and these plant male spores are a part of sexual plant reproduction. However, due to their microcapsular organization, it is possible to use them in pharmacy as drug transporters (Diego-Toboada et al. 2014). The external shells of such materials have many properties, such as good elasticity, physical and chemical resistance, UV shielding capability, and antioxidant activity. These properties enable the plant to protect its genetic material from external factors, such as oxidation by air, sunlight, and physical stress. This shell has two distinct layers: the inner layer, called the intine, which is composed largely of cellulose, and an external layer, called the exine, which consists mainly of a polymer known as sporopollenin. This polymer is composed of carbon, hydrogen, and oxygen, and although its exact structure remains unknown, it has been described as an oxidative polymer of carotenoids or polyunsaturated fatty acids as well as conjugated phenols.

The fluorescence of pollen could be useful in distinguishing between a self-compatible and a self-incompatible pollen of the same species. The pollen's viability and self-compatibility may be identified from their autofluorescence (Roshchina et al. 1998a, b; Kovaleva and Roshchina 1999; Roshchina 2003, 2008). Most significant are the maturing and fertility of pollen as the male gametophyte for the fertilization leading to the formation of fruits and seeds. The fluorescence of pollen grains is also useful in meteorological and environmental monitoring, especially for the determination of peaks in allergic hazards for humans. Moreover, pollen fluorescence is a useful indicator in paleontology and criminalistic and forensic medicine (van Gujzel 1961, 1967, 1971; Roshchina et al. 1997b).

As a drug, pollen is used in alternative, homoeopathic, and folk medicine, although for pharmacy using the vegetative parts of plants, contamination with airborne pollen also takes place. Some industries make various preparations from collected native pollen, from dried pollen packets to tablets. For example, special technology has been demonstrated by Xiamen Yiyu Biological Technology Co., Ltd. Fujian, China.

Also, bees may collect pollen (called *pollen load*) on their legs to feed their young and carry it into the beehive. This intact pollen is also used as a native drug preparation. Then, in the beehive, pollen is treated with bee saliva and mixed with nectar collected by worker bees. Similar bee products are also used in alternative medicine and in order to increase stamina and improve athletic ability. Other uses, not proved by research, have included premenstrual syndrome, premature aging, hay fever, nosebleeds, joint pain, painful urination, prostate problems, stomach problems, and other conditions (Williams 2013). Bee pollen is often sold as a herbal supplement. There are no regulated manufacturing standards in place for many herbal compounds, and some marketed supplements are contaminated with toxic metals or other drugs. Herbal/health supplements should be purchased from a reliable source to minimize the risk of contamination. Bee pollen and propolis are similar resinous substances collected from various plant sources by honeybees. Their biological properties include antioxidant, antimicrobial, antifungal, anti-inflammatory, and immunoregulatory actions. Propolis contains polyphenols and flavonoids, as well as several specific antioxidant compounds including beta-carotene, caffeic acid, and kaempferol (Williams 2013). Antimicrobial effects have been observed against Gram-positive bacteria and yeasts due to pollen's strong antioxidant activity resulting from its high total polyphenol content. A beneficial role of bee products (propolis, bee pollen, and royal jelly) has been found in the treatment of obesity and metabolic syndrome, as well as a reduction in cardiovascular risk factors, which are important for the early prevention of atherosclerosis, by various flavonoids (Handjiev et al. 2019). Pollen collected from different varieties of herbs can be used for directed therapeutics as well as the prevention of chronic diseases. Folk healers favor bee pollen to promote the activation of reproductive cells, restore sexual function, and create a favorable background for conception and child bearing. Bee pollen protects from everyday stress and overexertion, restores strength, and strengthens heart muscles, improving the work of the blood. Women will appreciate this product during menstruation and menopause. This natural phytopreparation is often given to humans to maintain and restore immunity. Essentially, this pollen contains almost all parts of the periodic table that are useful for the human body. The composition depends on the area where the bees collect pollen and the varieties of vegetation. Honey insects carefully select this product. It is easy to guess that the bee, following its natural instincts and the scent, may choose a product rich in benefits and power. Bee pollen is often given to children to maintain and restore immunity (for this, consultation is needed). Caring parents prefer to enhance their bodies not only during the virulence and season of colds, but in advance. This helps to protect against viral and bacterial infections. Bee pollen is also an excellent restorative tool, which will support and help the body cope with the consequences of catarrhal or other infectious disease.

5.3.2 Fluorescence of Intact Pollen and Pollen Products

The higher intensity of light emission in the male gametophyte, known as pollen, and the position of maxima in the fluorescence spectra differentiate these plant male cells from animal sexual cells (Roshchina et al. 1996, 1997b; Roshchina and Melnikova 1996, 1999). The fluorescence of 21 grass species changed (20–40 nm shifts to longer wavelengths) during the first 10–30 s after the UV light excitation was switched on (Driessen et al. 1989). In particular, within 30 s, the pollen cytoplasm of *Bromus hordeaceus* became yellow-orange instead of yellow. Pore emission in pollen grains of *Festuca arundinacea* and *Holcus mollis* was yellow, while in other species, it was white-blue. Among the 21 studied grass species, maxima in blue (460–490 nm) were specific to *Alopecurus pratensis*, *Arrhenatherum elatius*, *Dactilis glomerata*, *Elymus repens*, *Holcus lanatus*, *Lolium perenne*, *Molinea caerulea*, *Phalaris arundinacea*, *Phleum pratensis*, and *Poa trivialis*, while *Poa pratensis*, *Agrostis stolonifera*, and *Alopecurus pratensis* fluoresce in green (500–530 nm), and *Bromus hordeaceus*, *Cynosurus cristatus*, *Festuca arundinacea*, and *Holcus mollis* fluoresce in green-yellow or yellow (540–575 nm). The fluorescence of pollen may be useful for the estimation of the cell state and viability as well as self-incompatibility in the contact of pollen as the male gametophyte and the pistil as the female gametophyte (Roshchina et al. 1997b, Roshchina 2008).

The registration of the fluorescence spectra of pollen grains of some plants collected in natural conditions shows the variety of peaks related to the different components of their surface and excreta. Among them are mono-, di-, and three-component systems. Van Gijzel (1961, 1967, 1971), Willemse (1971), and Driessen with coworkers (1989) recorded pollen fluorescence spectra of forest trees and herbaceous plants and observed one or two maxima in them, mainly in the blue and orange-red regions. Usually, the fluorescence spectra of pollen have maxima at 460–490, 510–550, and 620–680 nm. The first could be connected with phenolic compounds, the second with carotenoids, and the third with chlorophyll when the pollen is immature (Wolfbeis 1985, Chappele et al., 1990; Roshchina 2008). Clearly seen maxima are found in the blue-green region at 480–500 nm (corn scabious, wild heliotrope, catch weed, hawthorn) and/or second maxima in the yellow-orange region (celandine poppy, delphinium, apple tree). Sometimes, a third maximum appears in the red region at 600–650 nm, such as for apple tree pollen grains. As shown in a monograph (Roshchina 2008), there are from one to three maxima at 460–490, 510–550, and 620–680 nm in pollen fluorescence spectra of various species. It depends on the substances included in the sporopollenin (a complex component in the surface cover, named exine) and/or exuded onto the pollen surface. Unlike fluorescence, the pollen absorbance spectra have from one to three maxima at 380–390, 410–450, and 760–840 nm (Roshchina 2008).

When pollen grains are mature, they often have no maxima in the red region of the fluorescence spectra, whereas in immature pollen grains of catch weed, a maximum at 680 nm, specific to chlorophyll, is observed. The fluorescence of dry immature and mature pollen differs. For instance, immature pollen of *Tussilago farfara* demonstrated green fluorescence with maxima at 465, 518, and 680 nm in the fluorescence spectra, whereas mature pollen, which lacks chlorophyll, showed mainly yellow fluorescence with maxima at 465, 520, and 535–540 nm (Roshchina et al. 1997a, 1998a). Blue fluorescence (maxima at 465–470 nm and a small shoulder at 665 nm) of immature pollen of *Philadelphus grandiflorus* changed to yellow-orange emission (maxima at 465, 510–520, and 620 nm) in mature pollen grains (Roshchina et al. 1997a, 1998a). Maxima in mature pollen occur due to the synthesis of carotenoids (maximum 520 nm) and azulenes (maximum 620–640 nm) during the development of the generative microspores (Roshchina et al. 1995, 1997a, 1998a). Luminescent and laser-scanning confocal microscopy have the potential to distinguish mature (without chlorophyll) and immature pollen fluorescing in red due to the chlorophyll. The fluorescence spectra of mature and immature pollen grains of greenhouse and outdoor plants have been shown in an earlier paper (Roshchina et al. 1998a). These differ in intensity of light emission and in the position of maxima of the fluorescence spectra. Mature pollen grains often have no 680 nm maximum, whereas in immature pollen grains of *Hymenocallis*, a 680 nm maximum, typical of chlorophyll, was observed. Pollen mainly demonstrates clearly seen maxima in the blue-green region (480–500 nm). Sometimes, a third maximum in the orange (550–570 nm) or red region (600–650 nm) appears, as for *Matricaria chamomilla*, for example. There are one to three maxima (at 460–490, 510–550, and 620–680 nm) in the pollen fluorescence spectra of various species. It depends on the substances contained in the sporopollenin or/and exuded onto the pollen surface. For instance, p-coumaric acid, a monomer in the sporopollenin skeleton, as in pollen of *Pinus mugo* Turra (Wehling et al. 1989), may emit in blue. Thus, the color of fluorescence, the fluorescence spectra, and the intensity of emission at the maxima may be useful for the analysis of mature and immature pollen or the readiness of pollen to germinate.

5.3.3　POLLEN IN ATMOSPHERE AS ECOLOGICAL AND ALLERGIC FACTOR

The pollen accumulation of some primary and secondary products can be released and is able to induce allergy. In some cases, foreign pollen pollutes collected samples of plant aboveground parts. Incident light fluorescence microscopy is growing rapidly in importance as an investigational tool in the fields of medical and biological research as well as field analysis of pollens.

The fluorescence of woody plants' pollens is used as a marker in airborne pollution (O'Connor et al. 2014). Pollen is a seasonal problem for meteorologists, who analyze weather changes, and for millions of people around the world who suffer from allergenic reactions to the antigens embedded on the outer casing of these microscopic grains (Knox 1979, 1984, Leuschner 1993). Tiny grains of pollen are released into the atmosphere by a wide spectrum of flowers, trees, weeds, grasses, and other plants that reproduce seasonally. Although most of these fertilizing gametes seldom reach their destinations, many find their way into the noses and throats of unsuspecting humans, where they elicit an allergic reaction. Commonly referred to as "hay fever," pollen allergies induce rhinitis and other effects that often cause sneezing, coughing, red and watery eyes, and related symptoms. An entire drug industry has arisen to deal with these unwanted medical effects, and for most people, a cure is as close as the nearest drug store or supermarket.

5.3.4 FLUORESCENT DIAGNOSTICS OF HONEY AND BEE PRODUCTS—POLLEN LOADS AND PERGA

The analysis of honey and other bee products on the basis of autofluorescence of their components is also useful (Roshchina et al. 1997c). Such components are pollen in honey, pollen load (pollen collected by bees), perga (transformed pollen load), and propolis (bee glue). As seen in Figure 5.8, the fluorescence spectra of native pollen collected by bees, pollen load, and bee products, the composition of the pollen differs among the various pollens.

They fluoresce in the blue–orange spectral region. The maxima are at 450–470 nm. The fluorescence of pollen loads and perga is due to pigments such as flavonoids and azulene, whereas maxima at 500–560 nm are due to carotenoids and coumarins. In some samples, both types of maxima are present. As for perga (pollen load which passes through the bees' nutrient system), it includes only transformed pigments that result in non-expressed maxima in their fluorescence spectra. The same picture is seen for propolis (bee glue) enriched with phenolic compounds. In propolis, flavonoids predominate in samples.

In some cases, depending on the season of the pollen load collection, we can determine possible pollens included in the pollen load. For instance, maximum at 520–530 nm is specific to *Trifolium repens*, *T. pretense*, and *Salix caprea*. Maximum 550 nm is seen in the fluorescence spectra of *Taraxacum officinale*, *Medicago falcata*, and *Cerasus vulgaris* (Roshchina 2008). The plants from which bees collect pollen and nectar for honey include meadow species such as clovers, dandelions, or willows flowering in earlier spring. Some of them have a similarity in their fluorescence spectra with pollen loads collected by bees (Figure 5.8). As demonstrated in Figure 5.8, pollen loads from samples were multicomponent, although there are predominant components among the fluorescence spectra. In the pollen load (1), the maximum was seen, mainly in green-yellowish (520–530 nm), while in other pollen loads (2, 3) at the same maximum was smoother. This peak usually correlated with an increased content of carotenoids (Roshchina et al. 1997c). In comparison with the fluorescence spectra of pollen received from clovers, both *Trifolium pratense* and *T. arvense*, we should note the same maximum at 520–530 nm, as is characteristic of meadow plants (Figure 5.8) as well as pollen from dandelion, *Taraxacum officinale* (Roshchina 2008). Immature pollen

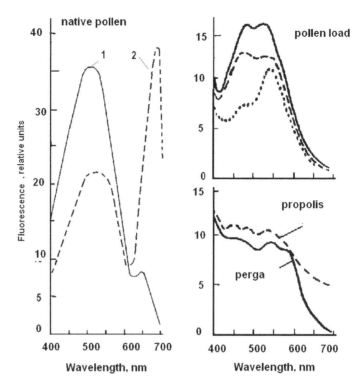

FIGURE 5.8   The fluorescence spectra of native pollen collected by bees in temporal cli-
matic zones, pollen loads, and products of bee transformation: perga and propolis. Native
pollen: (1) *Trifolium pratense*; (2) *Trifolium repens*. Pollen load: solid lines are the types
of pollen predominant in the sample. *Sources*: Roshchina et al. 1997c; Pharmacia (Russia)
3: 20–21; Roshchina, V.V., *Fluorescing World of Plant Secreting Cells*, Science Publishers,
Enfield, NH, 2008.

from clovers also has a maximum at 680 nm, specific to chlorophyll, whereas the
pollen loads studied do not. In the blue region of the spectra, pollen grains from
various samples of pollen loads fluoresce with maximum at 460–480 nm, and the
same peak is characteristic for willow, *Salix caprea*. Depending on the time when
the pollen load was collected by bees, from analysis of the fluorescence spectra or
the color of the fluorescence under a luminescence microscope, one can establish
what pollen prevailing maxima in the bee collection. Moreover, the maxima of
the predominant pollen grains indicate the main components; for example, green-
yellow indicates carotenoids, while blue – flavonoids and some terpenes (azulenes). Red fluorescence may show the presence of chlorophyll and in some cases,
azulenes. Thus, the autofluorescence of pollen loads may serve as an indicator in
express analysis of pollen loads without lengthy biochemical procedures. As for
the products of pollen, such as transformed products of bees (perga [pollen food]
or propolis [bee glue serving for the defense of the beehive against diseases]), their
fluorescence spectra have smoothed maxima and, therefore, are not as informative as the fluorescence spectra of pollen loads. Pollens are also found in various

samples of honey, and fluorescence spectra of the samples may give information about the plant species visited by bees and which nectar may be the basis of the honey sample.

Flavonoids such as kaempferol fluoresce in blue, and their deficit leads to a decrease in pollen fertility (Mo et al. 1992; Vogt et al. 1994). Several micromoles of kaempferol added to petunia pollen are enough to restore its normal fertility (Vogt et al. 1994). Pollination or wound-induced kaempferol accumulation in *Petunia* stigmas enhances seed production.

The cells of pollen and pistil may be viable or non-viable, and their fertility may be completely lacking after damage by some factor or the inability of some plant clones to self-pollinate. This knowledge is important for genetics and selection as well as for gene engineering. Autofluorescence of the cells could be informative in the analysis of fertility.

It is possible to assess cell viability by measuring the fluorescence of pollen moistened with nutrient medium. We can see changes in the fluorescence intensity of the untreated cells (autofluorescence) or of cells treated with the natural dye rutacridone (Roshchina 2002), as shown in Chapter 4. In both cases, the fluorescence is enhanced. Nevertheless, the intensity of autofluorescence increases to a larger degree, approximately threefold, in non-viable pollen (Table 5.2).

This fact was established not only for pollen of *Hippeastrum hybridum*, but also for pollen of *Philadelphus grandiflorus*, *Plantago major*, *Betula verrucosa*, and other species.

The fluorescence of pollen could be useful for distinguishing between a self-compatible and a self-incompatible pollen of the same species, as was seen for *Petunia* clones (Roshchina et al. 1997a; Kovaleva and Roshchina 1999). In self-incompatible clones, although pollen comes from the flower of the same species as the pistil, self-pollination is not impossible. In a study of the autofluorescence of the reproductive organs of *Petunia*, both self-compatible and self-incompatible clones, by microspectrofluorimetry, changes in the fluorescence spectra of pollen grains and the pistil stigma, related to the alterations in the composition of their surface components, have been observed. In the self-incompatible clone, there was no maximum at 620–640 nm in the fluorescence spectra, which correlated with the absence of azulenes. Moreover, the decrease of the fluorescence in the red spectral region correlated with the lack of azulene pigments for pollen grains and with a decrease in chlorophyll synthesis for the pistil stigma. The yellow fluorescence at 530–560

TABLE 5.2

Fluorescence Intensity (Relative Units) of Viable and Non-Viable Pollen of *Hippeastrum hybridum* after 30 min of Moistening

Pollen	Autofluorescence at 460–475 nm	Fluorescence at 530 nm after Treatment with Rutacridone
Viable	1.2 ± 0.1	1.23 ± 0.09
Non-viable	3.3 ± 0.2	1.7 ± 0.1

nm was also small due to the lower concentration of carotenoids (Roshchina et al. 1997a, 1998a).

Biotechnological applications to medicinal preparations from pollen cells are also possible. New approaches to the biotechnology of natural medicinal preparation include interactions with nanoparticles. Some fluorescence methods may be used in similar technology. For example, pollen undergoes adhesion to fullerene fine nanoparticles (Aoyagi and Ugwu 2011).

Spores of spore-bearing plants also may be transporters of drugs into human bodies. Besides their transport function, like a shell, spores are rich in many compounds that are useful for health. For example, spores formed on sporophytes of *Equisetum arvense* contain many flavonoids (Syrchina et al. 1980; Asgarpanah and Roohi 2012). Although direct experiments on animals are still only being planned, it is known based on data from thin-layer chromatography (Roshchina 2008) that spores including azulenes may show antiallergic effects.

5.4 SANITARY TREATMENT OF DRIED PHARMACEUTICAL PREPARATIONS BEFORE MARKETING

Before marketing dried phytopreparations such as herbs and parts of trees and shrubs, it is necessary to prepare them by sanitary treatment using ozonation or ultraviolet irradiation against parasites, microbial infections, and insects. Fluorescence an as indicator of possible influence may be also applied. Examples of similar testing for sanitary treatment by ozone or UV irradiation are shown in Table 5.3.

Treatment by ozone was tested for two dose variants: chronic exposure to a lower concentration for 3 days and acute concentration for 10 min. Among the species studied, the leaf autofluorescence of *Brachychiton diversifolius* and *Buddleja officinalis* was absent at both doses, while in *Nerium oleander*, there was a similar reaction at the highest dose of the gas. In contrast, in *Citrus unshiu* (no emission in control), emission appeared and strongly increased. The well-known medicinal plants *Ginkgo biloba* and *Eucalyptus cinerea* were especially sensitive to ozone at both dose variants. The maximum at 680–685 nm, specific to chlorophyll, decreased or disappeared first in most studied plants. Moreover, the reader can see that new peaks appeared on exposure to ozone. The most resistant to ozone was the maximum at 470 nm, specific to phenols, while shorter-wavelength peaks related to terpenoids disappeared or decreased.

Another treatment with UV irradiation also resulted in marked changes in autofluorescence (Table 5.3). Here, a release of ozone was in the conditions that confirms data, receved with pure ozone. The maximum at 470 nm was resistant to UV light or appeared under this exposure. Chlorophyll emission persisted only for *Ficus colchica*. In other species, one can see the lack of maxima lower than 450 nm. This correlates with the data on treatment with pure ozone.

It is necessary to keep these results in mind in the technology of preparation of dried pharmaceutical material. As well as chlorophyll, catecholamines may undergo oxidation (Allen 2003).

TABLE 5.3

Influence of Ozone or Ultraviolet Irradiation on Fluorescence of Dried Leaves with Pharmaceuticals. Excitation at 360/380 nm

	Fluorescence Maxima, nm			
Plant Species	**Control (without Treatment)**	**Ozone (Dose 0.05 μL/L for 3 Days)**	**Ozone (Dose 0.5 μl/L for 10 min)**	**UV Irradiation (Bactericidal Spectral Energy, watt/m²) 30 min. Analysis after 24 h**
Brachychiton diversifolius R.Br. (Malvaceae)	452, 470, 600, 685/452, 470, shoulder 633, 685	0/0	0/0	456, 470/0
Buddleja officinalis L. (Scrophulariaceae)	470, 680/470, 685	0/0	0/0	450
Citrus unshiu (Yu. Tanaka ex Swingle) Marcow (Rutaceae)	No emission	454, 479, small 690/454, 479, small 690 (high common emission)	470, small 610/470 ((high common emission)	460/460
Eucalyptus cinerea F. Muell. ex Benth. (Myrtaceae)	400, 413, 450, 470, 680/450, 470, 685	425, 450, 470, 675/454, 470, 680	430, 450, 470, 485, 490, small 680	454, 470/454, 470
Ficus colchica Grossh. (Moraceae)	460, 680/460, 680		450, 470, shoulder 530, 685/454, 470, 690 (common emission increases twofold)	460, 680/460, 680
Ginkgo biloba L. (Ginkgoaceae)	400, 410, 450, 470, 680/450, 470, 680	470, 600, 690/470, smallest 680	470, 630, smallest 690/454, 470, 680	0/0
Lantana camara L. (Verbenaceae)	470/470		454, 470, 685/470, 685 (common emission increases 10-fold)	470/470
Nerium oleander L. (Apocynaceae)	479/470, small 680	454, 470, 690/454, 470, 685	0/0	470/470

CONCLUSION

Interactions between various natural compounds in phytopreparations analyzed by fluorescence methods may be of interest for pharmacy, because they simultaneously model possible relations in intact secretory structures filled with drug substances. The purpose is to prepare more suitable extractions for tinctures, liniments, and oils. Fluorescence in modeling with drugs helps to have the maximal care effects from mixtures for the store nativity of the natural plant components, although few data are devoted to the problem yet. The fluorescent analysis of the pharmaceutical material after sanitary treatment enables the determination of optimal conditions for the procedure without missing useful characteristics for human health. Fluorescence is a possible indicator of the transformation of plant compounds.

Conclusion

Fluorescent methods for the analysis of both raw pharmaceutical materials and phytopreparations are at the beginning of their development. This monograph represents the first endeavor to provide a tool for pharmacologists based on the author's own experiments during the last almost 30 years. The first steps in the field were taken by biophysicists working with plant secretory cells, where a non-invasive luminescent technique was able to show the presence of some drugs on the surface or within secretory structures containing pharmaceuticals. The main fluorescence of secretory cells relates to secondary metabolites concentrated in extracellular space, vacuoles, and secretory vesicles as well as a certain amount of fluorescent secretions released. Today, researchers working in pharmacy may observe the natural fluorescence (autofluorescence) of plant material or use histochemical fluorimetry *in situ*.

Autofluorescence excited by ultraviolet (UV), violet, or blue light is specific to many plant secretions, which cause the phenomenon to occur in intact secretory cells. Therefore, analytical techniques based on autofluorescence monitoring can be utilized in order to obtain information. Any modification of the luminescent microscope—from the simple technique to microspectrofluorimetry and confocal microscopy—enables it to be seen. Non-invasive observation of the secretory cells, based on their autofluorescence, enables the living state of the structures to be analyzed. The observation of autofluorescence of plant secretory structures opened up new possibilities for express fluorescent analysis without lengthy biochemical procedures. Every new secretory cell may be a source of new discoveries when their autofluorescence and the chemical composition of the secretions are compared. Although our fundamental knowledge of autofluorescence may be thought to be insufficient, this phenomenon is of interest in the global problem of living cell luminescence. In this monograph, apparatuses already used in practical work with medicinal plants are presented in Chapter 1. Moreover, techniques that have potential for future usage are considered here. In any case, the autofluorescence or fluorescence of dyes may be successfully studied using the above-mentioned apparatuses.

We know little about the mechanisms of emission arising within the cell as well as its quenching. The emission is often quenched or stimulated by other components of the secretions. Therefore, specialists in biophysics, biochemistry, and physiology have the widest field of investigation as pioneers who apply fluorescent techniques to pharmaceuticals.

Up to now, the possibility of plant vegetative tissue fluorescing under UV or violet irradiation in a luminescence microscope was associated with chlorophyll. Meanwhile, secretory cells contain fluorescing substances, such as phenols and others, that make possible fluorimetric analysis of the intact cell *in vivo*. Various groups of substances have significant differences in their spectral features. The observed fluorescence of intact cells is the sum of the emissions of several different groups of compounds, either secreted out or accumulated within the cell and bound to the cellular surface. The biochemistry of components contained in secreting structures has not been well studied yet. However, compounds occurring in the secretions possess characteristic maxima in the absorbance and fluorescent spectra that are suitable for the identification of the compounds in mixtures.

The fluorescence spectra of intact secretory cells containing various substances differ in the maxima position and the intensity. This permits a preliminary discrimination of the predominant components in the secretory cell tested. As shown in Chapters 1 and 2, microspectrofluorimetry is especially useful for such studies. Modern confocal microscopes provide the possibility of registering the fluorescence spectra of microscopic objects by laser-scanning confocal microscopy, which may also have potential for plant secretory cells. In parallel with these intensively used techniques, new methods have appeared, including the ablation technology and smartphone modifications. They are developing, as noted in Chapter 1, and may be used for pharmacy in the future.

The visible fluorescence of a secreting cell depends on the chemotaxonomy and metabolism of secreted products stored within secretory structures as well as evacuated onto the surface of the secreting cell (see Chapters 1–3). The fluorescence position in the visible part of the spectrum for the main groups of secondary metabolites has been established in secretory structures. Most of them fluoresce in blue and only a few in green, yellow, and orange-red. Their emission differs from that of chlorophyll but is similar to that of reduced pyridine nucleotides and lipofuscins, which also emit in the blue and blue-green spectral regions, respectively.

The spectral characteristics of individual substances contained in secretory cells can be measured *in vivo*, which is significant for their identification. The observation of fluorescence makes it possible to identify secretory cells among non-secretory ones in all plant organs and to notice the accumulation of secretions on the surfaces of herbaceous and woody plants. Non-invasive fluorescent analysis of secreting cells and tissues may become one of the usual methods in all pharmacological botanical studies, education, and forensics, because there is no necessity to use artificial dyes for the examination of the plant material.

Autofluorescence is suitable in the analysis of secretory structures for photographing their internal and external images without histochemical staining and fixation. The images of cellular components appear to apply as botanical visual aids for education and handbooks for practical pharmacological laboratories and for teaching students. Moreover, the design of such images may have artistic value in the publication and promotion of biological knowledge.

The phenomenon of secretory cells' autofluorescence may be used for the analysis of changes in the surrounding medium, because the cells are sensitive to environmental shifts. Autofluorescence may serve as a biosensor and bioindicator reaction. It may apply in pharmacology for the analysis of a plant material without expensive biochemical procedures, based on the levels of the individual secretory products accumulated in similar cells.

Among factors that influence light emission is the physiological state of the fluorescing cell, organ, and whole organism. As seen in Chapters 2 and 5, the autofluorescence of intact secretory cells changed during cell development and due to various damaging factors. The filling of the cells with a secretion and its removal are easily observed in these conditions, which can serve as an indicator of the cell state *in vivo.* This reflects alterations in the composition or/and redox state of the secretory products accumulated. The phenomenon could be an indicator of the structures' formation, based on the appearance of their secretions at the earliest stages. The light emission of secreting structures may be useful for the analysis of plant maturing and aging or damage.

Today, there is real practical potential in the application of the autofluorescence phenomenon of secretory cells. The basis of the fluorescent method applied to pharmacy is the property of most biologically active compounds contained in secreting cells, especially with double bonds, to fluoresce under irradiation by UV or violet light. Some secretory products, such as sesquiterpene lactones and alkaloids, may selectively stain cellular compartments, which makes them potentially natural fluorescent dyes (Chapter 4). The ability to fluoresce under UV or violet light in luminescence microscopy is also useful in the search of new natural fluorescent dyes among fluorescent components of plant secretions.

One problem in the autofluorescence study of secretory cells is the effects of various factors—temperature, light intensity, the pH of the medium, the composition of the medium, the ability of external chemicals to oxidize or reduce the fluorescent substance, and others as noted in Chapter 2. Moreover, this new line in the study of the autofluorescence of plant secretory cells already has many applications in practice. Fluorescent analysis of plant secretory cells has potential for diagnostics due to the possibility of fast drug estimation in intact tissues and single cells without tissue homogenization procedures and long biochemical manipulations (see Chapter 1).

One potential non-invasive application is the observation of luminescence of plant materials in order to know whether or not they are ready for pharmacy, because the main valuable drugs are concentrated in secretory structures. The estimation of cell viability is also a potential use that is especially required for normal pollination and fertilization in genetic analysis for selection in agriculture and biotechnology. Moreover, there are some possibilities of using the fluorescence changes in the analysis of cell damage, which is necessary for ecological monitoring. Earlier diagnostics of the disturbances induced by oxidative stress, aging, or pest invasion may be based on the fluorescence of secretory products.

The application of fluorescent methods in histochemistry and cytochemistry enables the detection of certain medicinal compounds in secretory structures that contain drugs. In this case, as well as phenols and terpenes (predominantly seen with color reactions), scientists may see biogenic amines assayed after staining with reagents that induce fluorescence of the neurotransmitters. Fluorescent secondary metabolites encountered in secretory cells may be applied to histochemical staining; for example, some sesquiterpene lactones. Also, the compounds may be used in studies of the mechanisms of intercellular and intracellular interactions in any laboratory practice as fluorescent dyes and probes.

Interactions between various natural compounds in phytopreparations analyzed by fluorescence methods may be of interest for pharmacy, because they simultaneously model possible relations in intact secretory structures filled with drug substances. This will improve tinctures, liniments, and oils prepared for maximal care effects from mixtures for the store nativity of the natural plant components. In order to determine optimal conditions of the sanitary treatment of plant pharmaceutical materials against parasites and infections, fluorescent diagnostics is necessary for safety and nonhazardous storage of useful components.

The magic ray of actinic light in the luminescence microscope opens a new fluorescing world for every researcher in the field of pharmacy.

Although our book addresses the more specialized problems of medicine and pharmacy, the conflict between plants and humans should be noted (Pichersky 2018). Humans depend on plants not only as a food source but also for building and clothing materials and as sources of medicines, psychoactive substances, spices, pigments, and more. However, plants such as medicinal and food species that have existed for thousands of years in their natural habitat are suffering barbaric damage and destruction. All people need to defend our green brothers.

References

Acharya, A.S., Manning, J.M. 1983. Reaction of glycolaldehyde with proteins: latent cross-linking potential of alpha-hydroxyaldehydes. *Proceedings of the National Academy of Sciences of USA* 80(12): 3590–3594.

Achaya, K.T., Craig, B.M., Youngs, C.G. 1964. The component fatty acids and glycerides of castor oil. *Journal of the American Oil Chemists' Society* 41(12): 783–784.

Aliotta, G., Cafiero, G. 1999. Biological properties of rue (Ruta graveolens L.). Potential use in sustainable agricultural systems. In: *Principles and Practices in Plant Ecology. Allelochemical Interactions*, eds. K.M.M. Inderjit Dakshini and C.L. Foy, 551–563. Boca Raton, FL: CRC Press.

Allen, J.F. 2003. Superoxide as an obligatory, catalytic intermediate in photosynthetic reduction of oxygen by adrenaline and dopamine. *Antioxidants and Redox Signaling* 5(1): 7–14.

Almeida, A., Muzitano, V.F., Costa, S.S. 2006. 1-octen-3-O-a-L-arabinopyranosyl-(1–6)-b-glucopyranoside, a minor substance from the leaves of Kalanchoe pinnata (Crassulaceae). *Rev Bras Farmacogn* 16(4). João Pessoa Oct./Dec. 2006. doi:10.1590/S0102-695X2006000400008.

Al Mohammedi, H.I., Paray, B.A., Rather, I.A. 2017. Anticancer activity of EA1 extracted from Equisetum arvense. *Pakistan Journal of Pharmacy Science* 30(5): 1947–1950.

Al Snafi, A.E. 2016. Medical importance of Cichorium intybus: A review. *IOSR Journal of Pharmacy* 6(3): 41–56.

Al Snafi, A.E. 2017. The pharmacology of Equisetum arvense- A review. *IOSR Journal of Pharmacy* 7(2): 31–42.

Amer, M.S., Tawashi, R. 1991. Modified pollen grains for delivering biologically active substances to plants and animals. *U.S. Patent* 5,013,552.A, 7 May 1991.

Amer, M.S., Tawashi, R. 1994. Drug loaded pollen grains with an outer coating for pulsed delivery. *U.S. Patent* 5,275,819A, 4 January 1994.

Ames, R.N., Ingham, E.R., Reid, C.P.P. 1982. Ultraviolet-induced autofluorescence of arbuscular mycorrhizal root infections—An alternative method to clearing and staining methods for assessing infections. *Canadian Journal of Microbiology* 28(3): 351–355.

Andary, C., Mondolot-Cosson, L., Daï, G.H. 1996. In situ detection of polyphenols in plant-microorganism interactions. In: *Histology, Ultrastructure and Molecular Cytology of Plant-Microorganism Interactions*, eds. M. Nicole and V. Gianinazzi-Pearson, 43–54. Dordrecht: Kluwer Academic Press.

Aniszewski, T. 2007. *Alkaloids-Secrets of Life Alkaloid Chemistry, Biological Significance, Applications and Ecological Role*, 181 p. Amsterdam: Elsevier.

Anwar, F., Ahmad, N., Alkharfy, K., Gilani, A.U.H. 2015. Mugwort (Artemisia vulgaris) oils. In: *Essential Oils in Food Preservation, Flavor and Safety*, ed. V.R. Preedy, 573–579. Amsterdam: Elsevier IncAcademic Press.

Anya, A.O. 1966. The structure and chemical composition of the nematode cuticle. Observations on some oxyurids and Ascaris. *Parasitology* 56(1): 179–198.

Aoyagi, H., Ugwu, C.U. 2011. Fullerene fine particles adhere to pollen grains and affect their autofluorescence and germination. *Nanotechnology, Science and Applications* 4: 67–71.

Applewhite, P.B. 1973. Serotonin and norepinephrine in plant tissues. *Phytochemistry* 12(1): 191–192.

Arnao, M.B., Hernandez-Ruiz, J. 2019. The multi-regulatory properties of melatonin in plants. In: *Neurotransmitters in Plants: Perspectives and Applications*, eds. A. Ramakrishna and V.V. Roshchina, 71–101. Boca Raton, FL: CRC Press.

Asgarpanah, J., Roohi, E. 2012. Review Phytochemistry and pharmacological properties of Equisetum arvense. *Journal of Medicinal Plants Research* 6(21): 3689–3693.

Aslam, R., Bostan, N., Nabgha, A., Maleeha, M., Safdar, W. 2011. A critical review on halophytes: Salt tolerant plants. *Journal of Medicinal Plants Research* 5(33): 7108–7118.

Atkin, S.L., Beckett, S.T., Mackenzie, G. 2005. Dosage form comprising an exine coating of sporopollenin or derivatized sporopollenin. *WO Patent* 2005000280, 6 January 2005.

Atkin, S.L., Beckett, S.T., Mackenzie, G. 2007. Topical formulations containing sporopollenin. *WO Patent* 2007012857, 26 April 2007.

Atkinson, K.F., Kathem, S.H., Jin, X. et al. 2015. Dopaminergic signaling within the primary cilia in the renovascular system. *Frontiers in Physiology*. doi:10.3389/fphys.2015.00103.

Avissar, R., Menaché, R., Mandel, E.M., Rubinstein, I., Djaldetti, M. 1975. Increased salivary calcium levels as an indicator of digoxin intoxication. *Archives of Internal Medicine* 135(8): 1029–1032.

Axelsson, S., Björklund, A., Falk, B., Lindvall, O., Svensson, L.A. 1973. Glyoxylic acid condensation: A new fluorescence method for histochemical demonstration of biogenic monoamines. *Acta Physiologia Scandinavia* 87(1): 57–62.

Azizi, A., Wagner, C., Honermeier, B., Friedt, W. 2009a. Intraspecific diversity and relationships among subspecies of Origanum vulgare revealed by comparative AFLP and SAMPL marker analysis. *Plant Systematic and Evolution* 281(1–4): 151–160.

Azizi, A., Yan, F., Honermeier, B. 2009b. Herbage yield, essential oil content and composition of three oregano (Origanum vulgare L.) populations as affected by soil moisture regimes and nitrogen supply. *Industrial Crops and Products* 29(2–3): 554–561.

Badria, F.A., Aboelmaaty, W.S. 2019. Histochemistry: A versatile and indispensible tool in localization of gene expression, enzymes, cytokines, secondary metabolites and detection of plants infection and pollution. *Acta Scientific Pharmaceutical Sciences* 3(7): 88–100.

Bais, S., Abrol, N., Prashar, Y., Kumari, R. 2016. Modulatory effect of standardised amentoflavone isolated from Juniperus communis L. agianst Freund's adjuvant induced arthritis in rats (histopathological and X Ray anaysis). *Biomedicine and Pharmacotherapy* 86: 381–392.

Bais, S., Gill, N.S., Rana, N., Shandil, S. 2014. A phytopharmacological review on a medicinal plant: Juniperus communis. *International Scholarly Research Notices* 2014. doi:10.1155/2014/634723.

Bamel, K., Gupta, R. 2019. Acetylcholine as a regulator of differenciation and development in tomato. In: *Neurotransmitters in Plants: Perspectives and Applications*. eds. A. Ramakrishna and V.V. Roshchina, 113–132. Boca Raton, FL: CRC Press.

Barwell, C.J. 1979. The occurrence of histamine in the red alga of Furcellaria lumbricalis Lamour. *Botanica Marina* 22: 399–401.

Barwell, C.J. 1989. Distribution of histamine in the thallus Furcellaria lumbricalis. *Journal of Applied Phycology* 1(4): 341–344.

Bashirova, P.M., Amineva, A.A., Murtazina, F.K. 2003. Anthocyanins as criteria for the screening of azulene-enrich ed forms of Asiatic milfoil. In: *New and Non-Traditional Plants and Prospectss of their Utilization*. (Materials of 5 International Symposium. 9-14 June 2003, Pushchino.). 1 edition, eds. P.F. Kononkov and J.G. Millar, 110–112. Moscow: Russian University of International Friendship.

Bassolé, I.H.N., Juliani, H.R. 2012. Essential oils in combination and their antimicrobial properties. *Molecules* 17(4): 3989–4006.

Bauer, S., Stormer, E., Granbaum, H.J., Roots, I. 2001. Determination of hyperforin, and pseudohypericin in human plasma using high performance liquid chromatography analysis with fluorescence and ultraviolet detection. *Journal of Chromatography* 765 B: 29–35.

Beck, I., Jochner, S., Gilles, S., et al. 2013. High environmental ozone levels lead to enhanced allergenicity of birch pollen. *PLoS One* 8(11): e80147.

Berg, R.H. 2004. Evaluation of spectral imaging for plant cell analysis. *Journal of Microscopy* 214(2): 174–181.

Bergau, N., Bennewitz, S., Syrowatka, F., Hause, G., Tissier, A. 2015. The development of type VI glandular trichomes in the cultivated tomato Solanum lycopersicum and a related wild species S. habrochaites. *BMC Plant Biology* 15. doi:10.1186/s12870-015-0678-z.

Bergau, M., Santos, A.N., Henning, A., Balcke, G.U., Tissier, A. 2016. Autofluorescence as a signal to sort developing glandular trichomes by flow cytometry. *Frontiers in Plant Science* 7: 949.

Betz, J.M., Eppley, R.M., Taylor, W.C., Andrzejewski, D. 1994. Determination of pyrrolizidine alkaloids in commercial comfrey products (Symphytum sp.). *Journal of Pharmaceutical Sciences* 83(5): 649–653.

Bezuglov, V.V., Gretskaya, N.M., Esipov, S.S., et al. 2004. Fluorescent lipophilic analogs of serotonin, dopamine and acetylcholine: Synthesis, mass-spectrometry and biological activity. *Bioorganic Chemistry (Bioorg Khim, Russia)* 30(5): 512–519.

Bharti, S.K., Trivedi, A., Kumar, N. 2017. Air pollution tolerance index of plants growing near an industrial site. *Urban Climate.* doi:10.1016/j.uclim.2017.10.007.

Bhattacharjee, A. 2019. Phytoextracts and their derivatives affecting neurotransmission relevant to Alzheimer's disease: An ethno-medicinal perspective. In: *Neurotransmitters in Plants: Perspectives and Applications*, eds. A. Ramakrishna and V.V. Roshchina, 357–385. Boca Raton, FL: CRC Press.

Bhattacharjee, P., Chakraborty, S. 2019. Neurotransmitters in edible plants. Implications in human health. In: *Neurotransmitters in Plants: Perspectives and Applications*, eds. A. Ramakrishna and V.V. Roshchina, 387–407. Boca Raton, FL: CRC Press.

Big chemical encyclopedia. https://chempedia.info/index/ Chemical substances, components, reactions, process design some of essential oils, well known as acting on the respiratory apparatus.

Björklund, A., Lindvall, O., Svensson, L.A. 1972. Mechanisms of fluorophore formation in the histochemical glyoxylic acid method for monoamines. *Histochemie* 32(2): 113–131.

Blagoveshenskii, A.V. 1956. Biogenic stimulators and of Academy of Science biochemical nature of their action. *Bull of Fundamental Botanic Garden of USSR Academy of Sciences* 25: 3–20.

Blakkali, F., Averbec, F., Averbeck, D., Idaomar, M. 2008. Biological effects of essential oils–a review. *Food and Chemical Toxicology* 46: 446–475.

Bond, J., Donaldson, L., Hill, S., Hitchcock, K. 2008. Safranine fluorescent staining of wood cell walls. *Biotechnic and Histochemistry* 83: 3–4, 161–171.

Boppart, S.A., Herrmann, J., Pitris, C., Stamper, D.L., Brezinski, M.E., Fujimoto, J.G. 1999. High-resolution optical coherence tomography-guided laser ablation of surgical tissue. *Journal of Surgical Research* 82(2): 275–284.

Boschi, F., Fontanella, M., Calderan, L., Sbarbati, A. 2011. Luminescence and fluorescence of essential oils. Fluorescence imaging in vivo of wild chamomile oil. *European Journal of Histochemistry: EJH* 55(2). doi:10.4081/ejh.2011.e18.

Borg-Karlson, A.K., Valterova, I., Nilsson, L.A. 1994. Volatile compounds from flowers of six species in the family Apiaceae: Bouquets for different pollinators? *Phytochemistry* 35(1): 111–119.

Bosabalidis, A., Tsekos, I. 1982. Glandular scale development and essential oil secretion in Origanum dictamnus L. *Planta* 156(6): 496–504.

Bosabalidis, A.M. 2002. Structural features of Origanum species. In: *Oregano: The Genera Origanum and Lippia*, ed. S.E. Kintzios, 11–64. London: Taylor and Francis.

Bosabalidis, A.M., Kokkini, S. 1997. Infraspecific variation of leaf anatomy in Origanum vulgare grown wild in Greece. *Botanical Journal of the Linnean Society* 123(4): 353–362.

Braga, I., Pissolato, M.D., Souza, G.M. 2017. Mitigating effects of acetylcholine supply on soybean seed germination under osmotic stress. *Brazilian Journal of Botany* 40(3): 617–624.

Brockmann, H., Falkenhausen, E.H.F., Dorlars, A. 1950. Die Konstitution des hypericins. *Naturwissenschaften* 37(23): 540.

Brooks, R.I., Csallany, A.S. 1978. Effects of air,ozone, and nitrogen dioxide exposure on the oxidation of corn and soybean lipids. *Journal of Agricultural and Food Chemistry* 28(5): 1203–1209.

Buchachenko, A.A., Khloplyankina, M.S., Dobryakov, S.N. 1967. Electron-excited state quenching by organic radicals. *Optics and Spectroscopy (USSR)* 22(4): 554–559.

Buschmann, C., Langdorf, G., Lichtenthaler, H.K. 2000. Imaging of the blue, green and red fluorescence emission of plants: An overview. *Photosynthetica* 38(4): 483–491.

Buschmann, C., Lichtenthaler, H.K. 1998. Principles and characteristics of multi-colour fluorescence imaging of plants. *Journal of Plant Physiology* 152(2–3): 297–314.

Bystrov, N.S., Chernov, B.K., Dobrynin, V.N., Kolosov, M.N. 1975. The structure of hyperforin. *Tetrahedron Letters* 16(32): 2791–2794.

Bystrov, N.S., Gupta, Sh.R., Dobrynin, V.N., Kolosov, M.N., Chernov, B.K. 1976. Structure of the antibiotic hyperforin. *Doklady Akademii Nauk SSSR* 226(1): 88–90. PMID 1248360.

Calvert, J.G., Orlando, J.J., Stockwell, W.R., Wallington, T.J. 2015. *The Mechanisms of Reactions Influencing Atmospheric Ozone.* Oxford: Oxford University Press.

Cao, F., Li, D.X. 2009. Morphology-controlled synthesis of SiO2 hollow microspheres using pollen grain as a biotemplate. *Biomedical Materials.* doi:10.1088/1748-6041/4/2/025009.

Cardoso-Gustavson, P, Dias, M.G., Costa, F.OB., de Moura Leite Camargos, G., da Cruz Centeno, D. 2018. Imaging of glyphosate uptake and identification of early microscopic markers in leaves of C3 and C4 glyphosate-resistant and -susceptible species. *Ecotoxicology and Environmental Safety* 163: 502–513.

Carr, D.J., and Carr, G.M. 1970. Oil glands and ducts in Eucalyptus L.'Herit.II. Development and structure of oil glands in the embryo. *Austral. J. Bot.* 18: 191–212.

Carr, D.J., Carr, G.M. 1970. Oil glands and ducts in Eucalyptus L.'Herit.II. Development and structure of oil glands in the embryo. *Australian Journal of Botany* 18: 191–212.

Cassaniti, C., Romano, D., Flowers, T.J. 2012. The response of ornamental plants to saline irrigation water. In: *Irrigation-Water Management, Pollution and Alternative Strategies,* ed. I. Garcıa-Garizabal, 131–139. Croatia, London: InTechOpen.

de Castro Marilia, Diego Demarco 2008. Phenolic compounds produced by secretory structures in plants: A brief review. *Natural Product Communications* 3(8): 1205–1376.

Cathokleous, E.A., Aitani, C.J.S., Takayoshi, K.O. 2015. T ropospheric O3, the nightmare of wild plants: A review study. *Journal of Agricultural Meteorology* 71(2): 142–152.

Cerovic, Z.G., Samson, G., Morales, F., Tremblay, N., Moya, I. 1999. Ultraviolet-induced fluorescence for plant monitoring: Present state and prospects. *Agronomie, EDP Sciences* 19(7): 543–578.

Četojević-Simin, D.D., Čanadanović-Brunet, J.M., Bogdanović, G.M., et al. 2010. Antioxidative and antiproliferative activities of different horsetail (*Equisetum arvense* L.) extracts. *Journal of Medicinal Food* 13(2): 452–459.

Chadwick, L.R., Pauli, G.F., Farnsworth, N.R. 2006. The pharmacognosy of Humulus lupulus L. (hops) with an emphasis on estrogenic properties. *Phytomedicine* 13(1–2): 119–131.

Chance, B., Thorell, B. 1959. Localization and kinetics of reduced piridine nucleotide in living cells by microfluorimetry. *Journal of Biological Chemistry* 234(11): 3044–3050.

Chandra, K., Jain, S.K. 2016. Therapeutic potential of Cichorium intybus in life style disorders; a review. *Asian Journal of Pharmaceutical and Clinical Research* 9(3): 20–25.

Chappele, E.W., McMurtrey, J.E., Kim, M.S. 1990. Laser induced blue fluorescence in vegetation. In: *Remote Sensing Science for the Nineties IGARSS'90: 10th Annu. Int. Geosci. and Remote Sensing Symp*, Washington, DC, 20–24 May 1990, ed. R. Mills, vol. 3, 199–122. College Park, MD: Univ. Maryland. The Institute of Electrical and Electronic Engineers, INC.

Chappelle, E.W., Wood, P.M., McMuitrey, J.E., Newcourt, W.W. 1985. Laser-induced fluorescence of green plants. 3: LIF spectral signatures of five major plant types. *Applied Optics* 24(1): 74–80.

Chatterjee, S.S., Bhattacharya, S.K., Wonnemann, M., Singer, A., Müller, W.E. 1998. Hyperforin as a possible antidepressant component of Hypericum extracts. *Life Sciences* 63(6): 499–510.

Cheryl, L. 2007. Ethnomedicines used in Trinidad and Tobago for reproductive problems. *Journal of Ethnobiology and Ethnomedicine* 3: 13.

Christensen, J., Nørgaard, L., Bro, R., Engelsen, S.B. 2006. Multivariate autofluorescence of intact food systems. *Chemical Reviews* 106(6): 1979–1994.

Cohen, E. 1987. Chitin biochemistry—Synthesis and inhibition. *Annual Review of Entomology* 32(1): 71–93.

Combrinck, S., Du Plooy, G.W., McCrindle, R.L., Botha, B.M. 2007. Morphology and Histochemistry of the Glandular Trichomes of Lippia scaberrima (Verbenaceae). *Annals of Botany* 99(6): 1111–1119.

Corcel, M., Devaux, M.-F., Guillon, F., Barron, C. 2016. Comparison of UV and visible autofluorescence of wheat grain tissues in macroscopic images of cross-sections and particles. *Computers and Electronics in Agriculture* 127: 281–288.

Coté, H., Boucher, M.A., Pichette, A., Legault, J. 2017. Anti-inflammatory, antioxidant, antibiotic, and cytotoxic activities of *Tanacetum vulgare* L. essential oil and its constituents. *Medicines (Basel)* 4(2): 34.

Croateau, R., Leblanc, R.M. 1977. Colchicine fluorescence measured with a laser spectrofluorimeter. *Journal of Luminescence* 15(3): 353–356.

Cross, S.A.M., Even, S.W.B., Rost, F.W.D. 1971. A study of the methods available for the cytochemical localization of histamine by fluorescence induced with o–phthalaldehyde or acetaldehyde. *Histochemical Journal* 3(6): 471–476.

Cseke, L.J., Setzer, W.N., Vogler, B., Kirakosyan, A., Kaufman, P.B. 2006. Traditional, analytical and preparative separations of Natural products. In: *Natural Products from Plants*, eds. L.J. Cseke et al. 105–173, 2- edition. Boca Raton, FL: CRC Press.

Czymmek, K.J., Fogg, M., Powell, D.H. 2007. In vivo time-lapse documentation using confocal and multi-photon microscopy reveals the mechanisms of invasion into the Arabidopsis root vascular system by *Fusarium oxysporum. Fungal Genetics and Biology* 44(10): 1011–1023.

Dayan, F.D., Zaccaro, M.L.M. 2012. Chlorophyll fluorescence as a marker for herbicide mechanisms of action. *Pesticide Biochemistry and Physiology* 102(3): 189–197.

derwich, Elhoussine, Chabir, R., Taouil, R., Senhaji, O. 2011. In-vitro antioxidant activity and GC/MS studies on the leaves of Mentha piperita (Lamiaceae) from Morocco. *International Journal of Pharmaceutical Sciences and Drug Research* 3(2): 130–136.

da Silva Baptista, M., Bastos, E.L. 2019. Fluorescence in pharmautics and cosmetics. In: *Fluorescence in Industry*, Springer Ser Fluoresc, eds. B. Pedras. Switzerland: Springer Nature AG. doi:10.1007/4243_2018_1.

Denk, W., Strickler, J.H., Webb, W.W. 1990. Two-photon laser scanning fluorescence microscopy. *Science* 248(4951): 73–76.

Devrnja, N., Anđelković, B., Aranđelović, S., Radulović, S., Soković, M., Krstić-Milošević, D., Ristić, M., Ćalić, D. 2017. Comparative studies on the antimicrobial and cytotoxic activities of *Tanacetum vulgare* L. essential oil and methanol extracts. *South African Journal of Botany* 111: 212–221.

De Sousa, F.D., Rogenio, F., Rogenio da Silva Mendes, F., et al. 2019. Plant macromolecules as biomaterials for wound healing. In: *Wound Healing*, London: Intech Open. doi:10.5772/Intechopen.89105.

de Vega, G.B., Ceballos, J.A., Anzalone, A., Gratton, E. 2016. A laser-scanning confocal microscopy study of carrageenan in red algae from seaweed farms near the Caribbean entrance of the Panama Canal. *Journal of Applied Phycology* 29(1): 495–508.

Diego-Toboada, A., Beckett, S.T., Atkin, S.L, Mackenzie, G. 2014. Hollow pollen shells to enhance drug delivery. *Pharmaceutics* 6(1): 80–96.

Diego-Taboada, A., Maillet, L., Banoub, J.H., et al. 2013. Protein free microcapsules obtained from plant spores as a model for drug delivery: Ibuprofen encapsulation, release and taste masking. *Journal of Materials Chemistry* 1: 707–713.

Di Lorenzo, G., Leto-Barone, M.S., La Piana, S., Di Bona, D. 2017. Unmet needs in understanding sublingual immunotherapy to grass pollen. In: *Immunotherapy - Myths, Reality, Ideas, Future*, ed. Kr. Metodiev. London: IntechOpen, 26 April 2017. doi:10.5772/67212. https://www.intechopen.com/books/immunotherapy-myths-reality-ideas-future/unmet-needs-in-understanding-sublingual-immunotherapy-to-grass-pollen.

Dobretsov, G.E., Syrejschikova, T.I., Smolina, N.V. 2011. On mechanisms of fluorescence quenching by water. *Biophysics (Russia)* 59(2): 231–237.

Dobson, H.E.M., Groth, I., Bergströmm, G. 1996. Pollen advertisement: Chemical contrasts between whole-flower and pollen odors. *American Journal of Botany* 83(7): 877–885.

Donaldson, L., Williams, N. 2018. Imaging and spectroscopy of natural fluorophores in pine needles. *Plants (Basel)* 7(1): pii: E10.

Dos Santos, J.G. Jr., Blanco, M.M., Do Monte, F.H., et al. 2005. Sedative and anticonvulsant effects of hydroalcoholic extract of *Equisetum arvense*. *Fitoterapia* 76(6): 508–513.

Douabalé, S.E., Dione, M., Coly, A., Tine, A. 2003. Contributions to the determination of histamine rate by measuring out the histamine-orthophthalaldehyde complex in the absorbtion and fluorescence. *Talanta* 60(2–3): 581–590.

Drabent, R., Pliszka, B., Olszewska, T. 1999. Fluorescence properties of plant anthocyanin pigments. I. Fluorescence of anthocyanins in Brassica oleracea L. extracts. *Journal of Photochemistry and Photobiology, Part B: Biology* 50(1): 53–58.

Driessen, M.N.B.M., Willemse, M.T.M., van Luijn, J.A.G. 1989. Grass pollen grain determination by light- and ultra-violet microscopy. *Grana* 28(2): 115–122.

Duke, J.A. 2002. *Handbook of Medicinal Herbs*. 2nd edition. Boca Raton, FL: CRC Press.

Duke, S.O., Paul, R.N.Jr.I., Lee, M. 1988. Terpenoids from genus Artemisia as potencial pesticides. In: *Biologically Active Natural Products*, ed. H.G. Cutler, 318–334. Washington, DC: American Chemical Society.

Ebadi, M. 2007. *Pharmacodynamic Basis of Herbal Medicine*. 2nd edition. Boca Raton, FL: CRC Press, 699 p.

Efremov, A.P. 2014. *Medicinal Plants and Fungi of Middle Lands of Russia. Complete Atlas-determinant*. Moscow: PhytonXXI.

Eftink, M.R. 1991. Fluorescence quenching: Theory and applicatins. In: *Topics in Fluorescence Spectroscopy*, ed. J.R. Lakowicz, vol. 2, pp. 53–126. New York and London: Plenum Press.

Eftink, M.R., Ghiron, C.A. 1981. Fluorescence quenching studies with proteins. *Analytical Biochemistry* 114(2): 199–227.

Egorov, I.A., Egofarova, R.Kh. 1971. The study of essential oils of blossom pollen of grape. *Doklady AN SSSR* 199: 1439–1442.

Ekici, K., Coskun, H. 2002. Histamine content of some commercial vegetable pickles. In: *Proceedings of ICNP-2002 – Trabzon*, Turkey, 162–164.

Ekici, K., Omer, A.K. 2019. Biogenic amines in plant food. In: *Neurotransmitters in Plants: Perspectives and Applications*, eds. A. Ramakrishna and V.V. Roshchina, 305–330. Boca Raton, FL: CRC Press.

Elakovich, S.D. 1988. Terpenoids as models for new agrochemicals. In: *Biologically Active Natural Products*, ed. H.G. Cutler, 250–261. Washington, DC: American Chemical Society.

Elist, J. 2006. Effects of pollen extract preparation prostat/poltit on lower urinary tract symptoms in patients with chronic nonbacterial prostatitis/chronic pelvic pain syndrome: A randomized, double-blind, placebo-controlled study. *Urology* 67(1): 60–63.

Elshafie, H.S., Camele, I. 2017. An overview of the biological effects of some Mediterranean essential oils on human health. *BioMed Research International* 2017: 1–14.

Emmelin, N., Feldberg, W. 1947. The mechanism of the sting of the common nettle (*Urtica urens*). *Journal of Physiology* 106(4): 440–455.

Eninger, B., Håkanson, R., Owman, C., Sporrong, B. 1968. Histochemical demonstration of histamine in paraffin sections by a fluorescent method. *Biochemical Pharmacology* 17(9): 1997–1998.

Erels, B., Reznicek, G., Gökhan, S., et al. 2012. Antimicrobial and antioxidant properties of *Artemisia* L. species from western Anatolia. *Turkish Journal of Biology* 36: 75–84.

Erland, L.A.E., Saxena, P.K. 2017. Beyond a neurotransmitter: The role of serotonin in plants. *NeuroTransmitter* 4: e11538.

Erland, L.A.E., Saxena, P.K. 2019. Melatonin and serotonin in plant morphogenesis and development. In: *Neurotransmitters in Plants: Perspectives and Applications*, eds. A. Ramakrishna and V.V. Roshchina, 57–70. Boca Raton, FL: CRC Press.

Erland, L.A.E., Turi, C.E., Saxena, P.K. 2016. Serotonin: An ancient molecule and an important regulator of plant processes. *Biotechnology Advances* 34(8): 1347–1361.

Ermolaev, V.A., Bodunov, E.N., Sveshnikova, E.B., Shakhverdov, T.A. 1977. *Non-Radiating (Emitted) Transfer of Energy from Electron Excitation*. Leningrad: Nauka, 378 p.

Ernst, E. ed. 2003. *Hypericum: The Genus Hypericum*. Boca Raton, FL: CRC Press, 284 p.

Fahn, A. 1979. *Secretory Tissue in Plants*. London: Academic Press, 302 p.

Falasca, A., Caprari, C., De Felice, V., Iorizzi, M. 2016. GC-MS analysis of the essential oils of Juniperus communis L. berries growing wild in the Molise region: Seasonal variability and in vitro antifungal activity. *Biochemical Systematics and Ecology* 69: 166–175.

Falck, B. 1962. Observation on the possibilities of the cellular localization of monoamines by a fluorescence method. *Acta Physiologica Scandinavica* 56(Suppl 197): 5–25.

Fang, E.F., Ng, T.B. 2013. *Antitumor Potential and Other Emerging Medicinal Properties of Natural Compounds*. 2nd edition. Heidelberg: Springer.

Faugeras, G.M., Debelmas, J., Paris, R.R.M. 1967. Sur la presence et la repartition d'aminophenols (tyramine, 3-hydroxy-tyramine ou dopamine, et nor-adrenaline), chez l'Aconit Napel: *Aconitum napellus* L. *Comptes Rends des Academy Science*, 264: 1864–1867.

Feng, P.C., Haynes, L.I., Magnus, K.E. 1961. High concentration of (-) noradrenaline in *Portulaca oleracea* L. *Nature* 191: 1108.

Fennelly, M.J., Sewell, G., Prentice, M., O'Connor, D.J., Sodeau, J.R. 2018. Review: The use of real-time fluorescence instrumentation to monitor ambient primary biological aerosol particles (PBAP). *Atmosphere* 9(1). doi:10.3390/atmos9010001.

Fernandes, F.F., Cardoso-Gustavson, P., Alves, E.S. 2016. Synergism between ozone and light stress: Structurel responses of polyphenols in a woody Brazilian species. *Chemosphere* 155: 573–582.

Fernández-Riva, M., Bolhaar, S., Gonsalez-Mancebo, E. et al. 2006. Apple allergy across Europe: How allergen sensitization profiles determine the clinical expression of allergies to plant foods. *Journal of Allergy and Clinical Immunology* 118(2): 481–488.

Figueiredo, P., Pina, F., Vilas-Boas, L., Macanita, A.L. 1990. Fluorescence spectra and decays of malvadin 3,5-diglucoside in aqueous solutions. *Journal of Photochemistry and Photobiology, Part A: Chemistry* 52(3): 411–424.

Filatov, V.P. 1955. *Tissue Therapy (Teaching on Biogenic Stimulants)*. Odessa: Odessa Publishing House.

Fischer, N.H. 1991. Sesquiterpene lactones. In: *Methods in Plant Biochemistry*, eds. B.V. Charlewood and D. Banthorpe, vol. 7, 187–212. London: Academic Press.

Fischer, N.H., Weidenhamer, J.D., Bradow, J.M. 1989. Inhibition and promotion of germination by several sesquiterpenes. *Journal of Chemical Ecology* 15(6): 1785–1793.

Fluck, R.A., Jaffe, M.J. 1974. Cholinesterases from plant tissues VI. Distribution and subcellular localization in *Phaseolus aureus* Roxb. *Plant Physiology* 53(5): 752–758.

Fluck, R.A., Leber, P.A., Lieser, J.D., et al. 2000. Choline conjugates of auxins. I. Direct evidence for the hydrolysis of choline-auxin conjugates by pea cholinesterase. *Plant Physiology and Biochemistry* 38(4): 301–308.

Fluorimetry of edible oils. 2015. Powered by WordPress | Designed by Physics Open Lab. 18 December 2015. https://physicsopenlab.org/feed/ https://www.facebook.com/lodovico. lappetito https://physicsopenlab.org/2015/12/18/fluorimetry-of-edible-oils/.

Feijó, J.A., Moreno, N. 2004. Imaging plant cells by two-photon excitation. *Protoplasma* 223(1): 1–32.

Foy, J.M., Parratt, J.R. 1960. A note on the presence of noradrenaline and 5-hydroxytryptamine in plantain (*Musa sapientum, var. paradisiaca*). *Journal of Pharmacy and Pharmacology* 12: 360–364.

Francis, W.R., Powers, M.L., Haddock, S.H. 2014. Characterization of an anthraquinone Fluorescence from the bioluminescent, Pelagic Polychaete Tomopteris. *Luminescence: the Journal of Biological and Chemical Luminescence* 29(8): 1135.

Franke, R., Schilcher, H. eds. 2005. *Chamomile, Industrial Profiles*. Boca Raton, FL: CRC Press, 304 p.

Gahan, P.B. 1984. *Plant Histochemistry and Cytochemistry*. London and New York: Acad. Press. 301 p.

Galindo-Castañeda, T., Brown, K.M., Lynch, J.P. 2018. Reduced root cortical burden improves growth and grain yield under low phosphorus availability in maize. *Plant, Cell and Environment* 41(7): 1579–1592.

Gandía-Herrero, F., Escribano, J., García-Carmona, F. 2005b. Hetaxanthins as pigments responsible for visible fluorescence in flowers. *Planta* 222(4): 586–593.

Gandía-Herrero, F., García-Carmona, F., Escribano, J. 2005a. Fluorescent pigments: New perspectives in betalain research and applications. *Food Res Intl* 38(8–9): 879–884.

Gange, A.C., Bower, E., Stagg, P.G., Aplin, D.M., Gillam, A.E., Bracken, M. 1999. A comparison of visualization techniques for recording arbuscular mycorrhizal colonization. *New Phytologist* 142(1): 123–132.

Gao, T., Zhao, Q. et al. 2018. Effects of Exogenous dopamine on the uptake, transport, and resorption of apple ionome under moderate drought. *Frontiers in Plant Science* 9: 755–761.

Garcia, C.S., Menti, C., Lambert, A.P., et al. 2016. Pharmacological perspectives from Brazilian *Salvia officinalis* (Lamiaceae): Antioxidant, and antitumor in mammalian cells. *Anais da Academia Brasileira de Ciencias* 88(1): 281–292.

García-Caparros, P., Llanderal, A., Pestana, M., Correia, P.J., Lao, M.T. 2017. Nutritional and physiological responses of the dicotyledonous halophyte Sarcocornia fruticosa to salinity. *Australian Journal of Botany* 65(7): 573–581.

García-Plazaola, J.I., Fernández-Marín, B., Duke, S.O., Hernández, A., López-Arbeloa, F., Becerril, J.M. 2015. Autofluorescence: Biological functions and technical applications. *Plant Science: An International Journal of Experimental Plant Biology* 236: 136–145.

Gaydou, E.M., Randriamiharisoa, R., Bianchini, J.-P. 1986. Composition of the essential oil of ylang-ylang (*Cananga odorata* Hook Fil. et Thomson forma genuina) from Madagascar. *Journal of Agricultural and Food Chemistry* 34(3): 481–487.

Gersbach, P.V., Wyllie, S.G., Sarafis, V. 2001. A new histochemical method for localization of the site of monoterpene phenol accumulation in plant secretory structures. *Annals of Botany* 88(4): 521–525.

Golovkin, B.N., Rudenskaya, R.N., Trofimova, I.A., Shreter, A.I. 2001. *Biologically Active Substances of Plant Origin*, 3 volumes. Moscow: Nauka.

Gomes, M.P., Juneau, P. 2016. Oxidative stress in duckweed (*Lemna minor L.*) induced by glyphosate: Is the mitochondrial electron transport chain a target of this herbicide? *Environmental Pollution* 218: 402–409.

Gomes, M.P., Le Manac'h, S.G., Henault-Ethier, L. et al. 2017. Glyphosate-Dependent Inhibition of Photosynthesis in Willow. *Frontiers in Plant Science* 8: 207.

Goodger, J.Q., Heskes, A.M., Mitchell, M.C., et al. 2010. Isolation of intact sub-dermal secretory cavities from Eucalyptus. *Plant Methods*, 6: 20. doi:www.plantmethods.com/content/6/1/20

Goodger, J.Q., Heskes, A.M., Mitchell, M.C., et al. 2010. Isolation of intact sub-dermal secretory cavities from Eucalyptus. *Plant Methods* 6: 20. www.plantmethods.com/content/6/1/20.

Goodger, J.Q., Seneratne, S.L., Nicolle, D., Woodrow, I.E. 2016. Foliar essential oil glands of Eucalyptus subgenus Eucalyptus (Myrtaceae) Are a rich source of flavonoids and related non-volatile constituents. *PLoS One*. 2016. doi:10.1371/journal.pone.0151432.

Goodwin, B., Gomez, W., Fang, I.J., Meredith, Y., Sandhage, K.H. 2013. Conversion of pollen particles into three-dimensional ceramic replicas tailored for multimodal adhesion. *Chemistry of Materials* 25(22): 4529–4536.

Goryunova, S.V. (1952). The application of the fluorescence microscopy for the assay of living and dead cells of algae. *Trudy of the Institute of Microbiology of USSR Academy of Sciences* 2: 64–77.

Guan, Z.-S., Zhang, Y., Lu, C.-H., Xu, Z.-Z. 2008. Morphology-controlled synthesis of SiO2 hierarchical structures using pollen grains as templates. *Chinese Journal of Chemistry* 26(3): 467–470.

Guidotti, B.B., Gomes, B.R., Siqueira-Soares, R.D., Soares, A.R., Ferrarese-Filho, O. 2013. The effects of dopamine on root growth and enzyme activity in soybean seedlings. *Plant Signaling and Behavior* 8(9): e25477.

Guidry, G. 1999. A method for Counterstaining tissues in conjunction with the glyoxylic acid condensation reaction for detection of biogenic amines. *Journal of Histochemistry and Cytochemistry: Official Journal of the Histochemistry Society* 47(2): 261–264.

Gupta, B., Huang, B. 2014. Mechanism of salinity tolerance in plants: Physiological, biochemical and molecular characterization. *International Journal of Genomics* 2014: 19. Article ID 701596.

Gutteridge, J.M. 1995. Signal messenger and trigger molecules from free radical reactions and their control by antioxidants. In: *Signalling Mechanisms – From Transcription Factors to Oxidative* Stress, eds. L.P. Packer and K.W.A, 157–164. Wirtz. Heidelberg, Berlin: Springer in cooperation with NATO Sci. Affairs Division.

Hajjaj, G, Bahlouli, A., Tajani, M. et al. 2018. Analgesic and anti-inflammatory effects of *Papaver Rhoeas* L. A traditional medicinal plant of Morocco. *Journal of Natural and Ayurvedic Medicine* 2(7). doi: 10.23880/jonam-16000150

Hall, B., Lanba, A., Lynch, J.P. 2019. Three-dimensional analysis of biological systems via a novel laser ablation technique. *Journal of Laser Applications* 31(2): 1–5.

Hall, S.R., Bolger, H., Mann, S. 2003. Morphosynthesis of complex inorganic forms using pollen grain templates. *Chemical Communications*: 2784–2785.

Halliwell, B., Gutteridge, J.M.C. 1985. *Free Radicals in Biology and Medicine.* Oxford: Clarendon Press.

Hamidpour, M., Hamidpour, R., Hamidpour, S., Shahlari, M. 2014. Chemistry, pharmacology, and medicinal property of Sage (*Salvia*) to prevent and cure illnesses such as obesity,diabetes, depression, dementia, lupus, autism, heart disease, and cancer. *Journal of Traditional and Complementary Medicine* 4(2): 82–88.

Handjiev, S., Handjieva-Daelenska, T., Kuzeva, A. 2019. The role of bee products in the prevention and treatment of cardiometabolic disorders: Clinico-pharmacological and dietary studies. In: *The Role of Functional Food Security in Global Health*, eds. R. Watson, R. Singh, T. Takahashi, 449–456. Amsterdam: Academic Press Elsevier. doi:B978-0-12-813148-0.00026-8.

Harborne, J.B. 1993. *Introduction to Ecological Biochemistry.* 4th edition. London and Sandiego: Academic Press, 318 p.

Hartmann, E., Gupta, R. 1989. Acetylcholine as a signaling system in plants. In: *Second Messengers in Plant Growth and Development*, eds. W.F. Boss and D.I. Morve, 257–287. New York: Allan Liss.

Haugland, R.P. 2000. *Handbook of Fluorescent Probes and Research Chemicals.* 7 ed. Leiden: Molecular probes, Eugene.

Haugland, R.P. 2001. *Handbook of Fluorescent Probes and Research Products.* 8th edition. Leiden: Molecular Probes.

Healy, D.A., Huffman, J.A., O'Connor, D.J., Pöhlker, C., Pöschl, U., Sodeau, J.R. 2014. Ambient measurements of biological aerosol particles near Killarney, Ireland: A comparison between real-time fluorescence and microscopy techniques. *Atmospheric Chemistry and Physics* 14(15): 8055–8069.

Heidinder, A., Rabensteiner, D.F., Rabensteiner, J., et al. 2017. Decreased viability and proliferation of Chang conjunctival epithelial cells after contact with ultraviolet-irradiated pollen. *Cutaneous and Ocular Toxicology* 26. doi:10.1080/15569527.2017.1414226.

Heinrich, G., Pfeifhofer, H.W., Stabentheiner, E., Sawidis, T. 2002. Glandular hairs of Sigesbeckia jorullensis Kunth (Asteraceae): Morphology, histochemistry and composition of essencial oil. *Annals of Botany* 89(4): 459–469.

Hernandez, V., Mora, F., Melendez, P. 2012. A study of medicinal plant species and their ethnomedicinal values in Caparo Barinas, Venezuela. *Emirates Journal of Food and Agriculture* 24(2): 128–132.

Heskes, A.M. 2014. *An Investigation of Eucalypt Foliar Oil Gland Chemistry.* Ph.D. Dissertation. School of Botany, The University of Melbourne.

Heskes, A.M., Goodger, J.Q.D., Tsegay, S., Quach, T., Williams, S.J., Woodrow, I.E. 2012a. Localization of oleuropeyl glucose esters and a flavanone to secretory cavities of Myrtaceae. *PLoS One* 7(7): e40856.

Heskes, A.M., Lincoln, C.N., Goodger, J.Q.D., Woodrow, I.E., Smith, T.A. 2012b. Multiphoton fluorescence lifetime imaging shows spatial segregation of secondary metabolites in Eucalyptus secretory cavities. *Journal of Microscopy* 247(1–2): 33–42.

Heslop-Harrison, J. ed. 2014. *Pollen: Development and Physiology.* Oxford and Waltham, MA: Butterworth- Heinemann.

Heuer, S., Richter, S., Metzger, J.W., Wray, V., Nimtz, M., Strack, D. 1994. Betacyanins from bracts of *Bougainvillea glabra*. *Phytochemistry* 37(3): 761–767.

Hjorth, M., Mondolot, L., Buatois, B., et al. 2006. An easy and rapid method using microscopy to determine herbicide effects in Poaceae weed species. *Pest Management Science* 62(6): 5515–5521.

Hossain, M.A., Canning, J., Cook, K., Jamalipour, A. 2015. Smartphone laser spatial profiler. *Optics Letters* 40(22): 5156–5159. doi:10.1364/OL.40.005156.

Hossain, M.A., Canning, J., Cook, K., Jamalipour, A. 2016. Optical fiber smartphone spectrometer. *Optics Letters* 41(10): 2237–2241.

Huffman, J.A., Pöhlker, C., Treutlein, B., Pöschl, U. 2010. *Bioaerosol Analysis by Online Fluorescence Detection and Fluorescence Microscopy.*

Huisen, Y., Xinfu, X. 2003. A study on fluorescence spectra of Schiff bases. *Chemical Journal of Internet* 5(2): 17–23.

Hüsnü, K., Baser, C., Buchbauer, G. eds. 2009. *Handbook of Essential Oils; Science, Technology, and Applications.* Boca Raton, FL: CRC Press, 353–547.

Hutzler, P., Fischbach, R., Heller, W., et al. 1998. Tissue localization of phenolic compounds in plants by confocal laser scanning microscopy. *Journal of Experimental Botany* 49(323): 953–956.

Hwang, J.-U., Song, W.-Y., Hong, D., et al. 2016. Plant ABC transporters enable many unique aspects of terrestrial plant's lifestyle. *Molecular Plant* 9(3): 338–355.

Iriel, A., Lagoio, M.G. 2010a. Implication of reflectance and fluorescence of Rhododendron indicum flowers in biosignaling. *Photobiochem Photobiol Sciences* 9(3): 342–348.

Iriel, A., Lagoio, M.G. 2010b. Is the flower fluorescence relevant in biocommunication? *Naturwissenschafften* 97(10): 915–924.

Ivanchenko, V.A., Grodzinskii, A.M., Cherevchenko, T.M., et al. 1989. *Phytoergononomics.* Kiev: Naukova Dumka.

Jabaji-Hare, S.H., Perumalla, C.J., Kendrick, W.B. 1984. Autofluorescence of vesicles, arbuscules, and intercellular hyphae of a vesicular-arbuscular fungus in leek (*Allium porrum*) roots. *Canadian Journal of Botany-Revue Canadienne De Botanique* 62(12): 2665–2669.

Jacobs, J.F., Koper, G.J.M., Ursem, W.N. 2007. UV protecting coatings: A botanical approach. *Progress in Organic Coatings* 58(2–3): 166–171.

Jaldappagari, S., Motohashi, N., Gangeenahalli, M.P., Naismith, J.H. 2008. Bioactive mechanism of interaction between anthocyanins and macromolecules like DNA and proteins. *Topics in Heterocyclic Chemistry* 15: 49–65.

Jamalipour, A., Hossain, A. 2019. Smartphone intensity fluorimeter. In: *Smartphone Instrumentations for Public Health Safety.* doi:10.1007/978-3-030-02095-8_2.

Jamme, F., Villette, S., Giuliani, A., Rouam, V. 2010. Synchrotron UV fluorescence microscopy uncovers new probes in cells and tissues. *Microscopy and Microanalysis* 16(5): 507–514.

Jamme, F., Kascakova, S., Villette, S., et al. 2013. Deep UV autofluorescence microscopy for cell biology and tissue histology. *Biologie Cellulaire* 105(7): 277–288.

Jantová, S., Hudec, R., Secretár, S., Kučerák, J., Melušová, M. 2014. Salvia officinalis L. extract and its new food antioxidant formulations induce apoptosis through mitochondrial/caspase pathway in leukemia L1210 cells. *Interdisciplinary Toxicology* 7(3): 146–153.

Jiao, X., Li, Y., Zhang, X., et al. 2019. Exogenous dopamine application promotes alkali tolerance of apple seedlings. *Plants* 8(12): 580.

Jiao, Y., Zuo, Y. 2009. Ultrasonic extraction and HPLC determination of anthraquinones, aloeemodine, emodine, rheine, chrysophanol and physcione, in roots of *Polygoni multiflori*. *Phytochemical Analysis: PCA* 20(4): 272–278.

Jiménez-Becker, S., Ramírez, M., Plaza, B.M. 2019. The influence of salinity on the vegetative growth,osmolytes and chloride concentration of four halophytic species. *Journal of Plant Nutrition.* doi:10.1080/01904167.2019.1648666.

Joseph, B., Sridhar, S., Sankarganesh, Justinraj, Edwin, Biby T. 2011. Rare medicinal plant-*Kalanchoe pinnata. Research Journal of Microbiology* 6(4): 322–327.

Judzentiene, A., Budiene, J. 2017. Chemical polymorphism of essential oils of Artemisia vulgaris growing wild in Lithuania. *Chemistry and Biodiversity* 15(2). doi:10.1002/cbdv.201700257.

Jun, C., Li-Jiang, X., Ya-Ming, X. 2001. Three new phenolic glycosides from the fertile sprouts of Equisetum arvense. *Acta Botanica Sinica* 43(2): 193–197.

Jung, F., Arnold, H., Schoetensack, W., Bruno, G. 1949. Zur pharmacologischen Prufung entzundungswidriger azulene. *Klinische Wochenschrift* 27(29/30): 511–512.

Kabailiene, M. 2010. *Formation of Pollen Spectra and Interpretation Methods.* Saarbrukken: LAP Lambert Acad.Pub.

Kadohama, N., Goh, T., Ohnishi, M., Fukaki, H., Mimura, T., Suzuki, Y. 2013. Sudden collapse of vacuoles in *Saintpaulia* sp. palisade cells induced by a rapid temperature decrease. *PLoS One.* doi:10.1371/journal.pone.0057259.

Kakariari, E., Georgalaki, M., Kalantzopoulos, G., Tsakalidou, E. 2000. Purification and characterization of an intracellular esterase from Propionibacterium freudenreichii ssp. freudenreichii ITG 14. *Le Lait* 80(5): 491–501.

Kamo, K.K., Mahlberg, P.G. 1984. Dopamine biosynthesis at different stages of plant development in *Papaver somniferum. Journal of Natural Products* 47(4): 682–686.

Kang, S., Kang, K., Lee, K., Back, K. 2007. Characterization of tryptamine 5-hydroxylase and serotonin synthesis in rice plants. *Plant Cell Reports* 26(11): 2009–2015

Kang, S.S., Cordell, G.A., Soejarto, D.D., Fong, H. 2004. Alkaloids and flavonoids from Ricinus communis. *Journal of Natural Products* 48(1): 155–156.

Karabin, M., Hudkova, T., Jelinek, L., Dostàlek, P. 2016. Biotransformations and biological activities of hop flavonoids. *Biotechnology Advances* 33(6 Pt 2): 1063–1090.

Kariotti, A., Bilia, A.R. 2010. Hypericins as potential leads to new therapeutics. *International Journal of Molecular Sciences* 11(2): 562–594.

Karnaukhov, V.N. 1972. Spectral methods for investigations of energy regulation in living cells. In: *Biophysic of Living Cell*, eds. G.M. Frank and V.N. Karnaukhov, 100–117 Pushchino: Institute of Biophysics.

Karnaukhov, V.N. 1973. On the nature and function of yellow ageing pigment lipofuscin. *Experimental Cell Research* 80: 479–483.

Karnaukhov, V.N. 1978. *Luminescent Analysis of Living Cell.* Moscow: Nauka.

Karnaukhov, V.N. 1988a. *Spectral Analysis of Cell in Ecology and Environmental Protection (Cell Biomonitoring).* Pushchino: Biol. Center of USSR Academy of Sciences.

Karnaukhov, V.N. 1988b. *Biological Functions of Carotenoids.* Moscow: Nauka.

Karnaukhov, V.N., Yashin, V.A., Kulakov, V.I., et al. 1982. Apparatus for investigation of fluorescence characteristics of microscopic objects. *USA Patent*, N4, 354, 114: 1 - 14. *Patent of England (1983)* 2.039.03 R5R.CHI.

Karnaukhov, V.N., Yashin, V.A., Kazantsev, A.P., Karnaukhova, N.A., Kulakov, V.I. 1987. Double-wave microfluorimeter-photometer based on standard attachment. *Tsitologia* (Cytology, USSR) 29: 113–116.

Karnaukhova, N.A., Sergievich, L.A., Karnaukhov, V.N. 2010. Application of microspectral analysis to study intracellular metabolism in single cells and cell systems. *Natural Science* 2(5): 444–449.

Kautsky, H. 1939. Quenching of luminescence by oxygen. *Transactions of the Faraday Society* 35: 216–219.

Kavalali, G.M. ed. 2004. *Urtica. The Genus Urtica.* Boca Raton, FL: CRC Press, 112 p.

Kawashima, K., Misawa, H., Moriwaki, Y., et al. 2007. Ubiquitous expression of acetylcholine and its biological functions in life forms without nervous systems. *Life Sciences* 80(24–25): 2206–2209.

Khalid, K.A., da Silva, J. Teixeira. 2012. Biology of *Calendula officinalis* Linn.: Focus on pharmacology, biological activities and agronomic practices. *Medical and Aromatic Plant Science and Biotechnology* 6(1): 12–27.

Khoddami, A., Wilkes, M.A., Roberts, T.H. 2013. Techniques for analysis of plant phenolic compounds. *Molecules* 18(2): 2328–2375.

Kimura, M. 1968. Fluorescence histochemical study on serotonin and catecholamine in some plants. *Japanese Journal of Pharmacology* 18(2): 162–168.

Kintzios, S.E. ed. 2001. *Sage. The Genus Salvia.* Boca Raton, FL: CRC Press, 289 p.

Kintzios, S.E. 2002a. Profile of the multifaceted prince of the herbs. In: *Oregano: The Genera Origanum and Lippia. Medicinal and Aromatic Plants, Industrial Profiles 25*, ed. S.E. Kintzios, 3–9. Boca Raton, FL: Taylor & Francis/CRC Press, USA.

Kintzios, S.E. 2002b. The biotechnology of Oregano (*Origanum* sp. and *Lippia* sp.). In: *Oregano: The Genera Origanum and Lippia. Medicinal and Aromatic Plants, Industrial Profiles 25*, ed. S.E. Kintzios, 236–242. Boca Raton, FL: Taylor & Francis/CRC Press.

Knox, R.B. 1979. *Pollen and Allergy.* (Studies in Biology. V.107.). London: Arnold.

Knox, R.B. (1984). Pollen-pistil interactions. In: *Cellular Interactions. Encyclopedia of Plant Physiology*, eds. H.F. Linskens and J. Heslop, vol. 17, 508–608. Heidelberg, Berlin: Springer.

Knox, R.B., Clarke, A., Harrison, S., Smith, P., Marchaloni, J.J. 1976. Cell recognition in plants: Determinants of the stigma surface and their pollen interactions. *Proceedings of National Academy of Sciences of USA* 73: 2788–2792.

Knudsen, R.B., Tollsten, L., Bergstrome, L.G. 1993. The chemical diversity of floral scents – A checklist of volatile compounds isolated by head-space technique. *Phytochemistry* 33(2): 253–280.

Kokate, C., Jalalpure, S.S., Hurakadle, P.J. 2011. *Textbook of Pharmaceutical Biotechnology.* New Delhi: Elsevier.

Kokkini, S. 1997. Taxonomy, diversity and distribution of *Origanum* species. In: *Oregano, 14 Proceedings of the IPGRI International Workshop*, ed. S. Padulosi, 2–12. Italy, Rome. In: Oregano. *Proceedings of the IPGRI International Workshop on Oregano*, 8–12 May 1996, CIHEAM: Valenzano (Bari), Italy. Institute of Plant Genetics and Crop Plant Research, Gatersleben/International Plant Genetic Resources Institute, Rome, Italy.

Kokkini, S., Vokou, D. 1989. Carvacrol-rich plants in Greece. *Flavour and Fragrance Journal* 4(1): 1–7.

Kokkini, S., Vokou, D., Karousou, R. 1991. Morphological and chemical variation of *Origanum vulgare* L. in Greece. *Botanica Chronica* 10: 337–346.

Kolesnikova, R.D., Chernodubov, A.I., Deryuzhkin, R.I. 1980. The composition of essential oils of some species belonging to genera *Pinus L.* and Cedrus L. *Rastitelnye Resursy* (Plant resources, USSR) 16: 108–112.

Kolodziejczyk-Czepas, J., Stochmal, A. 2017. Bufadienolides of Kalanchoe species: An overview of chemical structure, biological activity and prospects for pharmacological use. *Phytochemistry Reviews: Proceedings of the Phytochemical Society of Europe* 16(6): 1155–1171.

Kongbonga, Y.G.M., Ghalila, H., Onana, M.B., et al. 2011. Characterization of vegetable oils by fluorescence spectroscopy. *Food and Nutrition Sciences* 2(7): 692–699.

Konovalov, D.A. 1995. Natural azulenes. *Rastitelnye Resursi (Plant Resources) in Russian* 31: 101–132.

Konovalov, D.A. 2019. Neurotransmitters in medicinal plants. In: *Neurotransmitters in Plants: Perspectives and Applications*, eds. A. Ramakrishna and V.V. Roshchina, 331–356. Boca Raton, FL: CRC Press.

Konovalov, D.A., Starykh, V.V. 1996. Sesquiterpene lactones - Phytotoxic substances of plants. In: *Proceedings of the 1st All-Russian Conference on Botanical Resources*, ed. A.L. Budantsev, 201–202. Sankt-Peterburg: Botanical Institute of RAS.

Kopach, M., Kopac, C., Klachek, B. 1980. Luminescent properties of some flavonoids. *Journal of Organic Chemistry, USSR* 16(8): 1721–1725.

Korobova, L.N., Beletskii, Yu.D., Karnaukhova, T.B. 1988. The experiment of the selection of salt-tolerant forms of sunflower among the selection material based on the content of histamine in seeds. *Fiziologiya and Biokhimiya Kulturnikh Rastenii (USSR)* 20: 403–406.

Kovaleva, L.V., Roshchina, V.V. 1999. Indication of the S-gene (related to a self-imcompatibilirty) expression based on autofluorescence of the system pollen-pistil. In: *Abstracts of the 4th All-Russian Congress of Plant Physiology*, Moscow, 4–9 October 1999, 598. Moscow: Inst. of Plant Physiology RAS.

Kucharski, M. 2005. For use of fluorescence method identification of weeds resistant to photosynthesis inhibitor herbicides. *F Progress Plant Protection* 45(2): 825–827.

Kuklin, A.I., Conger, B.V. 1995. Catecholamines in plants. *Journal of Plant Growth Regulation* 14(2): 91–97.

Kulisic, T., Radoni, A., Katalinic, V., Milos, M. 2004. Use of different methods for testing antioxidative activity of oregano essential oil. *Food Chemistry* 85(4): 633–640.

Kulma, A., Szopa, J. 2007. Catecholamines are active compounds in plants. *Plant Science* 172(3): 433–440.

Kumar, S., Pandey, A.K. 2013. Chemistry and biological activities of flavonoids: An overview. *The Scientific World Journal* 2013. ID 162750. doi:10.1155/2013/162750.

Kumari, P., Panwari, S., Namita, T., et al. 2017. Pigment profiling of flower crops: A review. *Ecology, Environment and Conservation* 23(2): 1000–1008.

Kurilov, D., Kirichenko, E., Bidyukova, G., Olekhnovich, L., Luu Dam, K. 2009. Composition of the essential oil of introduced mint forms of *Mentha piperita* and *Mentha arvensis* species. *Doklady Biological Sciences* 429: 538–540.

Kurkin, V.F., Shmygareva, A.A., Ryazanova, T.K, San'kov. 2014. Quantitative determination of total anthracene derivatives in *Frangula* syrup preparation. *Pharmaceutical Chemistry Journal* 48: 467–469.

Kuruma, K., Hirai, E., Uchida, K., Kikuchi, J., Terui, Y. 1994. Quantification of histamine by postcolumn fluorescence detection high-performance liquid chromatography using Ortho-phthaldehyde in tetrahydrofuran and reaction mechanism. *Analytical Sciences* 10(2): 259–265.

Kutchan, T.M., Rush, M., Coscia, C.J. 1986. Subcellular localization of alkaloids and dopamine in different vacuolar compartment of *Papaver bracteatum*. *Plant Physiology* 81(1): 161–166.

Kwon, Y.H., Wells, K.S., Hoc, H.H.C. 1993. Fluorescence confocal microscopy: Applications in fungal cytology. *Mycologia* 85(5): 721–733.

Lachenmeier, D.W. 2007. Assesing the Authenticity of Absinthe Using Sensory Evaluation and HPTLC analysis of the Bitter Principle Absinthin. *Food Research International* 40(1): 167.

Lakowicz, J.R. 2006. *Principles of Fluorescence Spectroscopy*. 3rd edition. New York: Springer.

Lavie, G., Mazur, Y., Lavie, D., Meruelo, D. 1995. The chamical and biological properties of hypericin – A compound with a broad spectrum of biological activities. *Medicinal Research Reviews* 15(2): 111–119.

Law, C., Exley, C. 2011. New insight into silica deposition in horsetail (*Equisetum arvense*). *BMC Plant Biology* 29. doi:10.1186/1471-2229-11-112.

Lawrence, B.M. ed. 2007. *Mint: The Genus Mentha*. Boca Raton, FL: CRC Press, 576 p.

Lazarau, R., Heinrich, M. 2019. Herbal medicine: Who cares? The changing views on medicinal plants and their roles in British lifestyle. *Phytotherapy Research*. 2019: 1–12.

Lersten, N.R., Curtis, J.D. 1989. Foliar oil reservoir anatomy and distribution in *Solidago cana densis* (Asteraceae, tribe Astereae). *Nordic Journal of Botany* 9(3): 281–287.

Leuschner, R.M. 1993. Human biometeorology, Part II. *Experientia* 49(11): 931–942.

Levitt, J., Kuimova, M., Yahioglu, G., Chung, P., Suhling, K., Phillips, D. 2009. Membrane-bound molecular rotors measure viscosity in live cells via fluorescence lifetime imaging†. *Journal of Physical Chemistry C* 113(27). doi:10.1021/jp9013493.

Li, W.H., Cao, J.J., Lu, R., Che, C.X., Wei, Y.J. 2016. Study on fluorescence properties of flavanone and its hydroxyl derivatives. *Guang Pu Xue Yu Guang Pu Fen Xi* 36(4): 1007–1012. www.ncbi.nlm.nih.gov/pubmed/30048098.

Lima, J.C., Danesh, P., Figueiredo, P., Pina, F.S., Macanita, A.L. 1994. Excited states of antho-cyanins: The chalcone isomers of malvidin 3, 5-diglucoside. *Journal of Photochemistry and Photobiology,. Part A: Chemistry* 59(4): 412–418.

Lin, H.J., Lu, H.H., Liu, K.M., et al. 2015. Toward live cell imaging of dopamine neurotrans-mission with fluorescent neurotransmitter analogues. *Chemical Communications* 51: 14080–14083.

Lindvall, O., Björklund, A., Svensson, L.A. 1974. Fluorophore formation from catechol-amines and related compounds in the glyoxylic acid fluorescence histochemical method. *Histochemistry and Cell Biology* 39(3): 197–227.

Ling-Lee, M., Chilvers, G.A., Ashford, A.E. 1977. A histochemical study of phenolic materi-als in mycorrhizal and uninfected roots of Eucaliptus fastigata Deane and Maiden. *New Phytologist* 78(2): 313–328.

Lissoni, P., Rovelli, F., Brivio, F., et al. 2009. A randomized study of chemotherapy versus biochemotherapy with chemotherapy plus *Aloe arborescens* in patients with metastatic cancer. *In Vivo* 23(1): 171–175.

Liu, F., Pan, C., Drumm, P., Ang, C.Y. 2005. Liquid chromatography-mass spectrometry stud-ies of St. John's wort methanol extraction: Active constituents and their transformation. *Journal of Pharmaceutical and Biomedical Analysis* 37(2): 303–312. PMID 15708671.

London, E., Feigenson, G.W. 1981. Fluorescence quenching in model membranes. 1. Characterization of quenching caused by a spin-labelled phospholipids. *Biochemistry* 20(7): 1932–1938.

Lorenzo, S., Capodilupo, A., Bietti, M. 2008. A reassessment of the association between azu-lene and fullerene: Possible pitfalls in the determination of binding constants through fluorescence spectroscopy. *Chemical Communications (Camb)* 39: 4744–4746.

Lugan, R., Niogret, M.F., Leport, L., et al. 2010. Metabolome and water homeostasis analysis of Thellungiella salsuginea suggests that dehydration tolerance is a key response to osmotic stress in this halophyte. *The Plant Journal* 64(2): 215–229.

Luzarowski, M., Skirycz, A. 2019. Emerging strategies for the identification of protein-metabolite interactions. *Journal of Experimental Botany* 70(18): 4605–4618.

Lynn, E.V., Lehman, A., Cain, R. 1926. The volatile oil of *Ledum grelandicum. Journal of the American Pharmacists Association* 15(4): 263–265.

Lyte, M. 2010. Microbial endocrinology: *A personal journey.* In: *Microbial Endocrinology. Interkingdom Signaling in Infectious Disease and Health*, eds. M. Lyte and P.P.E. Freestone, 1–18. New York and Berlin: Springer-Verlag.

Lyte, M., Ernst, S. 1992. Catecholamine induced growth of gram-negative bacteria. *Life Sciences* 50(3): 203–212.

Macoy, D.M., Kim, W., Lee, S.Y., Kim, M.G. 2015. Biotic stress related functions of hydroxy-cinnamic acid amide in plants. *Journal of Plant Biology* 58(3): 156–163.

MacWilliam, I.C. 1961. Electrophoretic separation of hop resins and their transformation prod-ucts. *Journal of the Institute of Brewing* 444. https://doi.org/10.1002/j.2050-0416.1961. tb01823.x.

Maheshwary, S.C., Gupta, R., Gharyal, P.K. 1982. Cholinesterase in plants. In: *Recent Developments in Plant Sciences: S.M Sircar Memorial Volume*, ed. S.P. Sen, 145–160, New Delhi, India: Today and Tomorrows Printers and Publishers.

Manning, W.J., Feder, W.A. 1980. *Biomonitoring Air Pollutants with Plants.* London: Applied Science Publ.

Markova, L.N., Buznikov, G.A., Kovačević, N., Rakić, L., Salimova, N.B., Volina, E.V. 1985. Histochemical study of biogenic monoamines in early ("Prenervous") and late embryos of sea urchins. *International Journal of Developmental Neuroscience* 3(5): 493–495, 497–499.

Maroli, M., Vadell, M.V., Iglesias, A., Padula, P.J., Gómez Villafañe, I.E. 2015. Daily movements and microhabitat selection of hantavirus reservoirs and other Sigmodontinae rodent species that inhabit a protected Natural Area of Argentina. *EcoHealth.* doi:10.1007/s10393-015-10-38-z.

Marquardt, P., Vogg, G. 1952. Pharmakologische und Chemische Untersuchungen uber Wirkstoffe in Bienenpollen. *Arzneimittel-Forschung (Germany)* 21(353): 267–271.

Masci, V.L., Bernardini, S., Modesti, L., Ovidi, E., Tiezzi, A. 2019. Medicinal plants as a source of alkaloids. In: *Medically Important Plant Biomes: Source of Secondary Metabolites, Microorganisms for Sustainbility*, eds. A. Egamberdieva and A. Tiezzi, 85–113. Singapore: Springer Nature.

Mashkovskii, M.D. 2010. *Medicinal Drugs.* 15th edition. Moscow: Meditsina.

Masuda, H., Takenaka, Y., Yamaguchi, A., Nishikawa, S., Mizuno, H. 2006. A novel yellowish-green fluorescent protein from the marine copepod, *Chiridius poppei*, and its use as a reporter protein in HeLa cells. *Gene* 372: 18–25.

Melnichuk, M.D., Boiko, A.L., Grigoryuk, I.P., et al. 2014. *Hop (Humulus lupulus L.): From Cell to Plant: Atlas.* Kiev: NUBiP Ukraine.

Melnikova, E.V., Roshchina, V.V., Karnaukhov, V.N. 1997. Microspectrofluorimetry of pollen. *Biophysics (Russia)* 42(1): 226–233.

Melnychuk, M.D., Yakubenko, B.Y., Likhanov, A.F. 2013. Structure, development and function of the secretory system of *Humulus lupulus* L. (Cannabaceae). *Ukrainian Botanical Journal* 70(3): 342–350.

Merzlyak, M.N. 1988. Liposoluble fluorescent "aging pigments" in plants. In: *Lipofuscin –1987: State of the Art*, ed. I. Nagy, 451–452. Budapest and Amsterdam: Academ. Kiado, Elsevier.

Merzlyak, M.N. 1989. *Active Oxygen and Oxidative Processes in Membranes of Plant Cell.* Itogi Nauki i Tekhniki (Ser. Plant Physiology, Vol.6.) Moscow: VINITI, 168 p.

Merzlyak, M.N., Pogosyan, S.I. 1988. Oxygen Radicals and lipid peroxidation in plant cell. In *Oxygen Radicals in Chemistry, Biology and Medicine*, ed. I.B. Afanas'ev, 232–253. Riga: Medicinal Institute.

Merzlyak, M.N., Plakunova, O.V., Gostimsky, S.A., Rumyantseva, V.B., Kovak, K. 1984. Lipid peroxidation in and photodamage to a light-sensitive chlorescence pea mutant. *Physiologia Plantarum* 62(2): 329–334.

Merzlyak, M.N., Zhirov, B.K. 1990. *Free Radical Oxidation in Chloroplasts at the Ageing of Plants.* Itogi Nauki i Tekhniki, Ser.Biophysics, vol. 40, 101–125. Moscow: VINITI.

Microscopic Characterization of Botanical Medicines. 2011. eds. R. Upton, A.F. Graff, G. Jolliffe, R. Länger and E.M. Williamson. Boca Raton, FL: CRC Press, 800 p.

Mimica-Dukic, N., Simin, N., Cvejic, J., Jovin, E., Orcic, D., Bozin, B. 2008. Phenolic compounds in field horsetail (*Equisetum arvense* L.) as natural antioxidants. *Molecules* 13(7): 1455–1464.

Mitsumoto, K., Yabusaki, K., Aoyagi, H. 2009. Classification of pollen species using autofluorescence image 413 analysis. *Journal of Bioscience and Bioengineering* 107(1): 90–94.

Momin, M.A., Kondo, N., Kuramoto, M., Ogawa, Y., Yamamoto, K., Shigi, T. 2012. Investigation of excitation on wavelength for fluorescence emission of *citrus* peel based on UV-VIS spectra. *EAEF (Engineering in Agriculture, Environment and Food)* 5(4): 126–132.

Monici, M. 2005. Cell and tissue autofluorescence research and diagnostic applications. *Biotechnology Annual Review* 11: 227–256.

Morsy, N. 2017. Cardiac glycosides in medicinal plants. In: *Aromatic and Medicinal Plants – Back to Nature*, ed. H.A. El-Shemy. London: IntechOpen. doi:10.5772/65963. https://www.intechopen.com/books/aromatic-and-medicinal-plants-back-to-nature/cardiac-glycosides-in-medicinal-plants.

Mukherjee, S., Ramakrishna, A. 2019. Current advancement of serotonin analysis in Plants. Insights into qualitative and quantitative methodologies. In: *Neurotransmitters in Plants. Perspectives and Applications*, eds. A. Ramakrishna and V.V. Roshchina, 293–303. Boca Raton, FL: CRC Press.

Mueller, T.C., Moorman, T.B., Locke, M.A. 1992. Detection of herbicides using fluorescent spectroscopy. *Weed Science* 40(2): 270–274.

Mulabagal, V., Wang, H., Ngouajio, M., Nair, M.G. 2009. Characterization and quantification of health beneficial anthocyanins in leaf chicory (*Cichorium intybus*) varieties. *European Food Research and Technology* 230(1): 47–53.

Mumford, R.A., Lipke, H., Laufer, D.A., Feder, W.A. 1972. Ozone-induced changes in corn pollen. *Environmental Science and Technology* 6(5): 427–430.

Mundargi, R., Poytroz, M., Park, S., et al. 2015. *Lycopodium* Spores: A Naturally Manufactured, Superrobust Biomaterial for Drug Delivery. *Advanced Functional Materials* 26(4): 487–497.

Murav'eva, D.A., Samylina, I.A., Yakovlev, G.P. 2007. *Pharmacognosy*. Moscow: Medicina.

Muravnik, L.E. 2008. The trichomes of pericarp in *Juglans* (Juglandaceae): Scanning microscopy, fluorescent microscopy and histochemistry. *Cytology (Russia)* 50: 636–642.

Muravnik, L.E., Kostina, O., Shavarda, V.A.L. 2016. Glandular trichomes of *Tussilago farfara* (Senecioneae, Asteraceae). *Planta* 244(3): 737–752.

Murch, S.J. 2006. Neurotransmitters, neuroregulators and neurotoxins in plants. In: *Communication in Plant-Neuronal Aspects of Plant Life*, eds. F. Baluska, S. Mancuso and D. Volkmann, 137–151. Berlin: Springer.

Murch, S.J., Campbell, S.S.B., Saxena, P.K. 2001. The role of serotonin and melatonin in plant morphogenesis: Regulation of auxin-induced root organogenesis in in vitro-cultured explants of St. John's wort (*Hypericum perforatum* L.). *In Vitro Cellular and Developmental Biology – Plant* 37(6): 786–793.

Murch, S.J., Krishnaraj, S., Saxena, P.K. 2000. Tryptophan is a precursor for melatonin and serotonin biosynthesis in in vitro regenerated St. John's wort (*Hypericum perforatum* L. cv. Anthos) plants. *Plant Cell Reports* 19(7): 698–704.

Murch, S.J., Saxena, P.K. 2002. Melatonin: A potential regulator of plant growth and development? *In Vitro Cellular and Developmental Biology – Plant* 38(6): 531–536.

Murch, S.J., Saxena, P.K. 2004. Role of indoleamines in regulation of morphogenesis in in vitro cultures of St.John's wort (*Hypericum perforatum* L.). *Acta Horticulturae* 629: 425–432.

Muresan, M.L., Oniga, I., Georgescu, C., et al. 2014. *Botanical and phytochemical studies on Tanacetum vulgare L.* from Transylvania. *Acta Medica Transilvanica* 2(4): 145–147. https://www.degruyter.com/view/j/aucft.2015.19.issue-.

Murray, A.P., Faraoni, M.B., Castro, M.J., Alza, N.P., Cavallaro, V. 2013. Natural AChE inhibitors from plants and their contribution to Alzheimer's disease therapy. *Current Neuropharmacology* 11(4): 388–413.

Nachlas, M.M., Seligman, A.M. 1949. The histochemical demonstration of esterase. *Journal of the National Cancer Institute* 9(5–6): 415–425.

Nalbandyan, R.M. 1986. Copper-containing proteins of brain and their significance in etyology of schizophrenia. *Neurochemistry (USSR)* 5(1): 74–84.

Nandagopal, S., RanjithaKumari, B.D. 2007. Phytochemical and antibacterial studies of chicory (*Cichorium intybus* L.) - A multipurpose medicinal plant. *Advances in Biological Research* 1(1–2): 17–21.

Netolitzky, F. 1932. Hautgewebe. Die pflanzenhaare Gebruder Borntraeger. In: *Handbuch der pflanzen anatomy*. Bd.4, ed. K. Linsbauer.

Newall, C.A., Barnes, J., Anderson, L.R. 2002. *Herbal Medicines: A Guide for Health Care Professionals*. London: Pharmaceutical Press. ISBN 978-0-85369-474-8.

O'Connor, D.J., Healy, D.A., Hellebust, S., Buters, J.T.M.J., Sodeau, J.R. 2013. Using the WIBS-4 (waveband integrated bioaerosol sensor) technique for the on-line detection of pollen grains. *Aerosol Science and Technology* 48(4): 341–349.

O'Connor, D.J., Healy, D.A., Sodeau, J.R. 2015. A 1-month online monitoring campaign of ambient fungal spore concentrations in the harbour region of Cork, Ireland. *Aerobiologia* 31(3): 295–314.

O'Connor, D.J., Lovera, P., LIacopono, D., O'Riordan, A., Healy, D.A., Sodeau, J.R. 2014. Using spectral analysis and fluorescence lifetimes to discriminate between grass and tree pollen for aerobiological applications. *Analytical Methods.* doi:10/02/2014 07: 08:33: 07.

Oh, J.H., Ha, I.J., Lee, M.Y. et al. 2018. Identification and metabolite profiling of alkaloids in aaerial parts of *Papaver rhoeas* by liquid chromatography coupled with quadrupole time-of-flight tandem mass spectrometry. *Journal of Separation Science* 41(12): 2517–2527.

Oleskin, A.V., Shenderov, B.A. 2019. Production of neurochemicals by microorganisms. Implications for microbiota–plants interactivity. In: *Neurotransmitters in Plants. Perspectives and Applications*, eds. A. Ramakrishna and V.V. Roshchina, 271–279. Boca Raton, FL: CRC Press.

Olšovská, J., Bostikova, V., Dušek, M., et al. 2016. Humulus lupulus L (hops)- a valuable source of compounds with bioactive effects for future therapies. *Military Medical Science Letters* 85(1): 19–30.

Ornan, I., Kartal, M., Kan, Y., Sener, B. 2008. Activity of essential oils and individual components against acetyl- and butyrylcholinesterase. *ZNaturforsch* 63c: 547–553.

Ortega-Villasante, C., Burénb, S., Blázquez-Castroc, A., Barón-Solaa, Á., Hernándeza, L.E. 2018. Fluorescent in vivo imaging of reactive oxygen species and redox potential in plants. *Free Radical Biology and Medicine.* doi:10.1016/j.freeradbiomed.2018.04.005.

Owokotomo, O., Ekundayo, I.A., Abayomi, T.G., Chukwuka, A.V. 2015. In-vitro anti-cholinesterase activity of essential oil from four tropical medicinal plants. *Toxicology Reports* 2: 850–857.

Pan, Y.L. 2014. Detection and characterization of biological and other organic-carbon aerosol particles in atmosphere using fluorescence. *Journal of Quantitative Spectroscopy and Radiative Transfer* 150. doi:10.1016/j.jqsrt.2014.06.007.

Pan, Y.L., Hill, S.C., Pinnick, R.G., House, J.M., Flagan, R.C., Chang, R.K. 2011. Dual-excitation-wavelength fluorescence spectra and elastic scattering for differentiation of single airborne pollen and fungal particles. *Atmospheric Environment* 45: 1555–1563.

Pan, Y.L., Pinnick, R.G., Hill, S.C., Rosen, J.M., Chang, R.K. 2007. Single-particle laser-induced-fluorescence spectra of biological and other organic-carbon aerosols in the atmosphere: Measurements at New Haven, CT and Las Cruces, NM, USA. *Journal of Geophysical Research* 112: D24S19.

Pan, Y.L.J.D., Eversole, P. Kaye, Foote, V., et al. 2006. Bio-aerosol fluorescence. In: *Optics of Biological Particles (NATO Science Series II: Mathematics Physics and Chemistry)*, eds. A. Hoekstra, V. Maltsev and G. Videen, 63–164. New York: Springer.

Pan, Y.L.S.C., Hill, R.G., Pinnick, R.G, Huang, H., Bottiger, J.R., Chang, R.K. 2010. Fluorescence spectra of atmospheric aerosol particles measured using one or two excitation wavelengths: Comparison of classification schemes employing different emission and scattering results. *Optics Express* 18(12): 12436–12457.

Parvari, R., Pecht, I., Soreq, H. 1983. A microfluorometric assay for cholinesterases, suitable for multiple kinetic determinations of picomoles of released thiocholine. *Analytical Biochemistry* 133(2): 450–456.

Parvu, M., Vlase, L., Fodorpataki, L., et al. 2013. Chemical Composition of celandine (*Chelidonium majus* L.) Extract and its Effects on *Botrytis tulipae* (Lib.) Lind Fungus and the Tulip. *Notulae Botanicae Horti Agrobotanici Cluj-Napoca* 41(2): 414–426.

Pauli, A., Schilche, H. 2009. In vitro antimicrobial activities of essential oils monographed in the European pharmacopoeia. In: *Handbook of Essential Oils; Science, Technology, and Applications.* Chapter 12. eds. K. Hüsnü, C. Baser and G. Buchbauer, 353–547. Boca Raton, FL: CRC Press.

Paul, A., Ge, Y., Prakash, S., Shum-Tim, D. 2009. Microencapsulated stem cells for tissue repairing: Implications in cell-based myocardial therapy. *Regenerative Medicine* 4(5): 733–745.

Paunov, V.N., Mackenzie, G., Stoyanov, S.D. 2007. Sporopollenin micro-reactors for in situ preparation, encapsulation and targeted delivery of active components. *Journal of Materials Chemistry* 17(7): 609–612.

Pawley, J.B. ed. 1990. *Handbook of Biological Confocal Microscopy.* New York: Plenum.

Pawley, J.B., Centonze, V.E. 1994. Practical laser scanning confocal light microscopy: Obtaining optical performance from your instrument. In: *Cell Biology: A Laboratory Handbook*, ed. J. Cells, 44–64. San Diego: Academic Press.

Pawley, J., Pawley, J.B. 2006. *Handbook of Biological Confocal Microscopy.* Berlin, Heidelberg: Springer-Verlag.

Peer, W.A., Murphy, A.S. 2006. Flavonoids as signal molecules (Targets of flavonoids action). In: *Science of Flavonoids*, ed. E. Grotewold, 239–268. Berlin: Springer.

Pérez-Recalde, M., Ruiz Arias, I.E., Hermida, É.B. 2018. Could essential oils enhance bio-polymers performance for wound healing? A systematic review. *Phytomedicine* 38: 57–65.

Pheophillov, P.P. 1944. The connection of fluorescence of organic dyes with their chemical structures. *Doklady USSR AcadSci* 45(9): 587–590.

Pichersky, E. 2019. *Plants and Human Conflict.* Boca Raton, FL: CRC Press.

Piras, I.S., Haapanen, L., Napolioni, V., Sacco, R., Van de Water, J., Persico, A.M. 2014. Anti-brain antibodies are associated with more severe cognitive and behavioral profiles in Italian children with autism Spectrum Disorder. *Brain, Behavior, and Immunity* 38: 91–99.

Póczi, D., Böddi, B. 2010. Studies on laticifers and milk of greater celandine (Chelidonium majus L.) with Fluorescence imaging and fluorescence spectroscopic methods. *Acta Pharmaceutica Hungarica* 80(3): 95.

Pöhlker, C., Huffman, J.A., Pöschl, U. 2011. Autofluorescence of atmospheric bioaerosols – Fluorescent biomolecules and potential interferences. *Atmospheric Measurement Techniques Discussions* 4(5): 5857–5933.

Polewski, K., Slawińska, D. 1987. Fluorescence and phosphorescence of adrenolutin. *Physiological Chemistry and Physics and Medical NMR* 19(2): 117–124.

Ponchet, M., Martin-Tanguy, J., Marais, A., Martin, C. 1982. Hydrocinnamoyl acid amines and aromatic amines in the influorescence of some Araceae. *Phytochemistry* 21: 2865–2870.

Poobathy, R., Zakaria, R., Murugaiyah, V., Subramaniam, S. 2018. Autofluorescence study and selected cyanidin quantification in the Jewel orchids *Anoectochilus* sp. and *Ludisia discolor. PLoS One* 13(4): e0195642.

Protacio, C.M., Dai, Y.R., Lewis, E.F., Flores, H.E. 1992. Growth stimulation by catechol-amines in plant tissue/organ cultures. *Plant Physiology* 98(1): 89–96.

Raal, A., Orav, A., Gretchushnikova, T. 2014. Essential oil content and composition in *Tanacetum vulgare* L. herbs growing wild in Estonia. *Journal of Essential Oil-Bearing Plants JEOP* 17(4). doi:10.1080/0972060X.2014.958554.

Racz-Kotilla, E., Adam, S., Galin, D. 1968. Protective effect of certain antiulcer medicines associated with azulene on experimental gastric ulcer. *Revue Médicale (Targu-Mures)* 14: 331–334.

Radha, M.H., Laxmipriya, L.P. 2015. Evaluation of biological properties and clinical effectiveness of *Aloe vera*: A systematic review. *Journal of Traditional and Complementary Medicine* 5(1): 21–26.

Radole, S., Kotwal, S.D. 2014. Equisetum arvense: Ethanopharmacological and Phytochemical review with reference to osteoporosis. *International Journal of Pharmaceutical Science and Health Care* 1(4): 131–141.

Radulović, N., Stojanović, G., Palić, R. 2006. Composition and antimicrobial activity of *Equisetum arvense* L. essential oil. *Phytotherapy Research: PTR* 20(1): 85–88.

Rahimi-Madiseh, M., Lorigoini, Z., Zamani-gharaghhoshi, H., Rafieian-kopaei, M. 2017. Berberis vulgaris: Specifications and traditional uses. *Iranian Journal of Basic Medical Sciences* 20(5): 569–587.

Rakić, V.P., Ota, A.M., Skrt, M.A., et al. 2014. Investigation of fluorescence properties of cyanidin -β-glucopyranoside. *Hemijska Industrija* 69(2): 155–163.

Ramakrishna, A., Giridhar, P., Jobin, M., Paulose, C.S., Ravishankar, G.A. 2011. Indoleamines and calcium enhance somatic embryogenesis in *Coffea canephora* P ex Fr. *Plant Cell, Tissue and Organ Culture* 108(2): 267–278.

Ramakrishna, A., Giridhar, P., Ravishankar, G.A. 2009. Indoleamines and calcium channels influence morphogenesisin in vitro cultures of *Mimosa pudica* L. *Plant Signaling and Behavior* 4(12): 1136–1141.

Ramakrishna, A., Giridhar, P., Ravishankar, G.A. 2011. Phytoserotonin: A review. *Plant Signaling and Behavior* 6(6): 800–809.

Ramakrishna, A., Rao, M.V., Davis, K.R. 1999. Ozone-induced cell death occurs via two distinct mechanisms in *Arabidopsis*: The role of salicylic acid. *Plant Journal: For Cell and Molecular Biology* 17(6): 603–614.

Ramakrishna, A., Rao, M.V., Davis, K.R. 2001. The physiology of ozone induced cell death. *Planta* 213(5): 682–690.

Rao, M.V., Koch, J.R., Davis, K.R. 2000. Ozone: A tool for probing programmed cell death in plants. *Plant Molecular Biology* 44(3): 345–358.

Rao, U.M. Vattikuti, Ciddi, V. 2005. An overview on *Hypericum perforatum* Linn. *Natural Product Radiance* 4(5): 368–381.

Rastitelnye Resursy. (Plant Resources). 1985–1996. ed. A.L. Budantsev. Leningrad-Sankt-Peterburg: Nauka. Issue 1 1985-1993, Sankt-Peterburg: Mir i Sem'ya, 571 p.

Raut, Jic. S., Karuppayil, S.M. 2014. A status review on the medicinal properties of essential oils. *Industrial Crops and Products* 62: 250–264.

Ravindran, P.N. 2017. *The Encyclopedia of Herbs and Spices.* Oxfordshire and Boston, MA: GABI: 2 vol./ Ravindran P.N. – Boston: GABI.

Reichling, J., Beiderbeck, R. 1991. X. *Chamomilla recutita* (L.) Rauschert (Camomile): In vitro culture and the production of secondary metabolites. In: *Biotechnology in Agriculture and Forestry*, ed. Y.P.S. Bajaj, vol. 15. Medicinal and Aromatic Plants III. 156–175. Berlin, Hedelberg: Springer-Verlag.

Reicosky, D.A., Hanover, J.W. 1978. Physiological effects of surface waxes: Light reflectance for glaucous and nonglaucous *Picea pungens*. *Plant Physiology* 62(1): 101–104.

Rekka, E.A., Chrysselis, M., Siskou, I., Kourounakis, A.P. 2002. Synthesis of new azulene derivatives and study of their effect on lipid peroxidation and lipogenase activity. *Chemical and Pharmaceutical Bulletin* 50(7): 904–907.

Rekka, E.A., Kourounakis, A.P., Kourounakis, P.N. 1996. Investigation of the effect of chamazulene on lipid peroxidation and free radical processes. *Research Communications inmolecular Pathology and Pharmacology* 92(3): 361–364.

Ribachenko, A.T., Georgievskii, V.P. 1975. Fluorescence characteristics of oxytransformed flavonoids. *Doklady AN Ukraine SSR, Ser. B* 11: 1009–1111.

Rice, E.L. 1984. *Allelopathy.* New York, San Francisco and London: Academic Press.

van Riel, M., Hammans, J.K., van den Ven, M., Verwer, W., Levine, Y.K. 1983. Fluorescence excitation profiles of β-carotene in solution and in lipid/water mixtures. *Biochemical and Biophysical Research Communications* 113(1): 102–107.

Roberts, M.F. 1986. Paper latex and alkaloid storage vacuoles. In: *Plant Vacuoles*, ed. B. Marin, 513–528. New York and London: Plenum Press.

Roberts, M.F., McCarthy, D., Kutchan, T.M., and Coscia, C.J., 1983. Localization of enzymes and alkaloidal metabolites in Papaver latex. *Archives of Biochemistry and Biophysics* 222: 599.

Rodriguez, E., Towers, G.H.N., Mitchell, J.C. 1976. Biological activities of sesquiterpene lactones. *Phytochemistry* 15(11): 1573–1580.

Rohringer, R., Kim, W.K., Samborski, D.J., Howes, H.K. 1977. Calcofluor: An optical brightener for fluorescence microscopy of fungal plant parasites in leaves. *Phytopathology* 67(6): 808–810.

Rohloff, J., Mordal, R., Dragland, S. 2004. Chemotypical variation of tansy (Tanacetum vulgare L.) from 40 different locations in Norway. *Journal of Agriculture and Food Chemistry* 52(6): 1742–1748.

Roriz, C.L.R.C., Barros, L., Carvalho, A.M., Santos-Buelga, C., Ferreira, I.C.F.R. 2014. Pterospartum tridentatum, Gomphrena globosa and Cymbopogon citratus: A phytochemical study focused on antioxidant compounds. *Food Research International* 62: 684–693.

Roshchina, V.D. 1962. Biological activity testing of tissue extracts based on the yeast fermentation energy and lifting power. In: *Tissue Preparations in Livestock Farming (Proceeding of Scientific-Industrial Conference)*, ed. G.F. Bondarenko, 227–233. Kiev: State Publishing House of Agricultural Literature of Ukrain SSR.

Roshchina, V.D. 1964. Tissue preparations of V.P. Filatov and growth of some plants. In: *Proceedings of the Voronezh Department of Ull-Union Botanical Society*, 75–81. Voronezh: VGU.

Roshchina, V.D., Roshchina, V.V. 1989. *The Excretory Function of Higher Plants*. Moscow: Nauka, 214 p.

Roshchina, V.V. 1989. Biomediators in chloroplasts of higher plants. I. The interaction with photosynthetic membranes. *Photosynthetica* 23: 197–206.

Roshchina, V.V. 1990a. Biomediators in chloroplasts of higher plants. 2. The acetylcholinehydrolyzing proteins. *Photosynthetica* 24: 110–116.

Roshchina, V.V. 1990b. Biomediators in chloroplasts of higher plants. 3. Effect of dopamine on photochemical activity. *Photosynthetica* 24: 117–121.

Roshchina, V.V. 1990c. Biomediators in chloroplasts of higher plants. 4. Reception by photosyn- thetic membranes. *Photosynthetica* 24: 539–549.

Roshchina, V.V. 1991. *Biomediators in Plants. Acetylcholine and Biogenic Amines*. Pushchino: Biological Center of USSR Academy of Sciences, 194 p.

Roshchina, V.V. 1999. Mechanisms of cell-cell communication. In: *Allelopathy Update*, ed. S.S. Narwal, vol. 2, 3–25. Enfield, NH: Science Publishers.

Roshchina, V.V. 2001. *Neurotransmitters in Plant Life*. Enfield, NH: Science Publishers.

Roshchina, V.V. 2002. Rutacridone as fluorescent dye for the study of pollen. *Journal of Fluorescence* 12(2): 241–243.

Roshchina, V.V. 2003. Autofluorescence of plant secreting cells as a biosensor and bioindicator reaction. *Journal of Fluorescence* 13(5): 403–420.

Roshchina, V.V. 2004. Cellular models to study the allelopathic mechanisms. *Allelopathy Journal* 13(1): 3–16.

Roshchina, V.V. 2005a. Allelochemicals as fluorescent markers, dyes and probes. *Allelopathy Journal* 16(1): 31–46.

Roshchina, V.V. 2005b. Contractile proteins in chemical signal transduction in plant microspores. *Biol Bull Ser Biol* 3: 281–286.

Roshchina, V.V. 2006a. Chemosignaling in plant microspore cells. *Biology Bulletin ser. Biol* 33(4): 332–338.

Roshchina, V.V. 2006b. Plant microspores as biosensors. *Trends of Modern Biology* 126(3): 262–274.

Roshchina, V.V. 2007a. Luminescent cell analysis in allelopathy. In: *Cell Diagnostics. Images, Biophysical and Biochemical Processes in Allelopathy*, eds. V.V. Roshchina and S.S. Narwal, 103–115. Enfield, Jersey (USA): Plymouth: Science Publisher.

Roshchina, V.V. 2007b. Cellular models as biosensors. In: *Cell Diagnostics: Images Biophysical and1 Biochemical Processes in* Allelopathy, eds. V.V. Roshchina and S.S. Narwal, 5–22. Plymouth: Science Publisher.

Roshchina, V.V. 2008. *Fluorescing World of Plant Secreting Cells*. Enfield, NH: Science Publishers.

Roshchina, V.V. 2009. Effects of proteins oxidants and antioxidants on germination of plant microspores. *Allelopathy Journal* 23(1): 37–50.

Roshchina, V.V. 2010. Evolutionary considerations of neurotransmitters in microbial, plant and animal cells. In: *Microbial Endocrinology: Interkingdom Signaling in Infectious Disease and Health*, eds. M. Lyte and P.P.E. Freestone, 17–52. New York: Springer.

Roshchina, V.V. 2012. Vital autofluorescence: Application to the study of plant living cells. *International Journal of Spectroscopy*. ID 124672: 1–14. doi:10.1155/2012/124672.

Roshchina, V.V. 2014. *Model Systems to Study Excretory Function of Higher Plants*. Dordrecht, Netherlands: Springer.

Roshchina, V.V. 2016a. New tendency and perspectives in evolutionary considerations of neurotransmitters in microbial, plant, and animal cells. In: *Advances in Experimental Medicine and Biology*, 874. Series: Microbial Endocrinology: Interkingdom Signaling in Infectious Disease and Health, ed. M. Lyte, pp. 25–77. Chapter 2. New York, Heidelberg: Springer International Publishing.

Roshchina, V.V. 2016b. The fluorescence methods to study neurotransmitters (biomediators) in plant cells. *Journal of Fluorescence* 26(3): 1029–1043.

Roshchina, V.V. 2017. Serotonin in cells: Fluorescent tools and models to study. In: *Serotonin and Melatonin (Serotonin and Melatonin: Their Functional Role in Plants, Food, Phytomedicine, and Human Health*, eds. Gokare A. Ravishankar and Akula Ramakrishna, 47–60. Boca Raton, FL: CRC Press.

Roshchina, V.V. 2018. Cholinesterase in secreting cells and isolated organelles. *Biological Membranes (Russia)* 35(2): 143–149.

Roshchina, V.V. 2019a. Tools for microanalysis of the neurotransmitter location in plant cell. In: *Neurotransmitters in Plants: Perspectives and Applications*. eds. A. Ramakrishna and V.V. Roshchina, 135–146. Boca Raton, FL: CRC Press.

Roshchina, V.V. 2019b. Inter-and intracellular signaling in plant cells with participation of neurotransmitters (biomediators). In: *Neurotransmitters in Plants: Perspectives and Applications*, eds. A. Ramakrishna and V.V. Roshchina, 147–180. Boca Raton, FL: CRC Press.

Roshchina, V.V. 2019c. Possible role of biogenic amines in plant-animal relations. In: *Neurotransmitters in Plants: Perspectives and Applications*, eds. A. Ramakrishna and V.V. Roshchina, 281–289. Boca Raton, FL: CRC Press.

Roshchina, V.V., Bezuglov, V.V., Markova, L.N., et al. 2003. Interaction of living cells with fluorescent derivatives of biogenic amines. *Doklady. Biochemistry and Biophysics* 393: 346–349.

Roshchina, V.V., Karnaukhov, V.N. 1999. Changes in pollen autofluorescence induced by ozone. *Biologia Plantarum* 42(2): 273–278.

Roshchina, V.V., Karnaukhov, V.N. 2010. The fluorescence analysis of medicinal drugs' interaction with unicellular biosensors. *Pharmacia (Russia)* 3: 43–46.

Roshchina, V.V., Melnikova, E.V. 1995. Spectral analysis of intact secretory cells and excretions of plants. *Allelopathy Journal* 2(2): 179–188.

Roshchina, V.V., Melnikova, E.V. 1996. Microspectrofluorometry: A new technique to study pollen allelopathy. *Allelopathy Journal* 3(1): 51–58.

Roshchina, V.V., Melnikova, E.V. 1998a. Allelopathy and plant reproductive cells: Participation of acetylcholine and histamine in signalling in the interactions of pollen and pistil. *Allelopathy Journal* 5(2): 171–182.

Roshchina, V.V., Melnikova, E.V. 1998b. Pollen-pistil interaction: Response to chemical signals. *Biology Bulletin* 25: 557–563.

Roshchina, V.V., Melnikova, E.V. 1998c. Chemosensory reactions at the interaction pollen-pistil. *Biology Bulletin* 6: 678–685.

Roshchina, V.V., Melnikova, E.V. 1999. Microspectrofluorimetry of intact secreting cells, with applications to the study of allelopathy. In: *Principles and Practices in Plant Ecology: Allelochemical Interactions*, eds. K.M.M. Inderjit and C.L.F. Dakshini, 99–126. Boca Raton, FL: CRC Press.

Roshchina, V.V., Melnikova, E.V. 2001. Chemosensitivity of pollen to ozone and peroxides. *Russian Journal of Plant Physiology* 48(1): 74–83.

Roshchina, V.V., Melnikova, E.V., Kovaleva, L.V. 1996. Autofluorescence in system polle-pistil in Hippeastrum hybridum. *Doklady RAS* 349(1): 118–120.

Roshchina, V.V., Melnikova, E., Karnaukhov, V.N., Golovkin, B.N. 1997a. Application of microspectrofluorimetry in spectral analysis of plant secretory cells. *Biological Bulletin (Russia)* 2: 167–171.

Roshchina, V.V., Melnikova, E.V., Kovaleva, L.V. 1997b. The changes in the fluorescence during the development of male gametophyte. *Russian Plant Physiology* 47(1): 45–53.

Roshchina, V.V., Melnikova, E.V., Spiridonov, N.A. 1997c. Spectral analysis of pollen load, perga and propolis. Pharmacia (Russia) 46(3): 20–22.

Roshchina, V.V., Melnikova, E.V., Kovaleva, L.V., Spiridonov, N.A. 1994. Cholinesterase of pollen grains. *Doklady Biological Sciences* 337: 424–427.

Roshchina, V.V., Melnikova, E.V., Mitkovskaya, L.I., Karnaukhov, V.N. 1998a. Microspectrofluorimetry for the study of intact plant secretory cells. *Journal Genetics Biology (Russia)* 59(5): 531–554.

Roshchina, V.V., Melnikova, E.V., Gordon, R.Ya., Konovalov, D.A., Kuzin, A.M. 1998b. A study of the radioprotective activity of proazulenes using a chemosensory model of *Hippeastrum hybridum* pollen. *Doklady Biophysics* 358–360: 20–23.

Roshchina, V.V., Melnikova, E.V., Spiridonov, N.A., Kovaleva, L.V. 1995. Azulenes, the blue pigments of pollen. *Doklady Biological Sciences* 340(1): 93–96.

Roshchina, V.V., Melnikova, E.V., Yashin, V.A., Karnaukhov, V.N. 2002. Autofluorescence of intact spores of horsetail Equisetum arvense L. during their development. *Biophysics (Russia)* 47(2): 318–324.

Roshchina, V.V., Roshchina, V.D. 2003. *Ozone and Plant Cell*. Dordrecht, the Netherlands: Kluwer Academic Publishers, 240 p.

Roshchina, V.V., Roshchina, V.D. 2012. *The Excretory Function of Higher Plants*. Saarbrü'cken: Lambert Academic Publishing, 466 p.

Roshchina, V.V., Yashin, V.A. 2013. Secreting cells of *Saintpaulia* as models in the study of plant cholinergic system. *Biological Membranes (Russia)* 23(6): 454–461.

Roshchina, V.V., Yashin, V.A. 2014. Neurotransmitters catecholamines and histamine in allelopathy: Plant cells as models in fluorescence microscopy. *Allelopathy Journal* 34(1): 1–16.

Roshchina, V.V., Yashin, V.A., Kononov, A.V. 2004. Autofluorescence of plant microspores studi ed by confocal microscopy and microspectrofluorimetry. *Journal of Fluorescence* 14(6): 745–750.

Roshchina, V.V., Yashin, V.A., Kuchin, A.V. 2015a. Microfluorescent analysis for bioindication of ozone on unicellular models. *Physics of Wave Phenomena* 23(3): 1–7.

Roshchina, V.V., Yashin, V.A., Kuchin, A.V. 2015b. Fluorescent analysis for bioindication of ozone on unicellular models. *Journal of Fluorescence* 25(3): 595 601.

Roshchina, V.V., Yashin, V.A., Kuchin, A.V. 2016a. Fluorescence of neurotransmitters and their reception in plant cell. *Biochemistry (Moscow), Supplement Series A: Membrane and Cell Biology* 10(3): 233–239. (Biological Membranes 33 (1),).

Roshchina, V.V., Kuchin, A.V., Yashin, V.A. 2016b. Autofluorescence of plant secretory cells as possible tool for pharmacy. *International Journal of Pharmacy and Chemistry* 2(2): 31–38.

Roshchina, V.V., Prizova, N.K., Khaibulaeva, L.M. 2019. Allelopathy experiments with Cha algae model. Histochemical analysis of the participation of neurotransmitter systems in water inhabitation. *Allelopathy Journal* 46(1): 17–24.

Roshchina, V.V., Shvirst, N.E., Kuchin, A.V.V. 2017a. The Autofluorescence Response of Flower Cells from *Saintpaulia ionantha* as the Biosensor Reaction to Ozone. *Computational Biology and Bioinformatics* 4(6): 60–66. Published online: Mar. 23, 2017.

Roshchina, V.V., Kuchin, A.V., Yashin, V.A. 2017b. Application of autofluorescence for analysis of medicinal plants. *Hindawi International Journal of Spectroscopy* 2017. Article ID 7159609, 8 pages.

Roshchina, V.V., Yashin, V.A., Kononov, A.V., Yashina, A.V. 2007. Laser-scanning confocal microscopy (LSCM): Study of plant secretory cells. In: *Cell Diagnostics. Images, Biophysical and Biochemical Processes in Allelopathy*, eds. V.V. Roshchina and S.S. Narwal, 93–102. Enfield, Jersey, Plymouth: Science Publisher.

Roshchina, V.V., Yashin, V.A., Vikhlyantsev, I.M. 2012. Fluorescence of plant microspores as biosensors. *Biochemistry (Moscow), Supplement SeriesA: Membrane and Cell Biology* 6(1): 105–112. Biological Membranes 28 (6): 547–556.

Roshchina, V.V., Yashin, V.A., Yashina, A.V., Goltyaev, M.V. 2011a. Colored allelochemicals in the modeling of cell-cellallelopathyc interactions. *Allelopathy Journal* 28: 1–12.

Roshchina, V.V., Yashin, V.A., Kuchin, A.V., Kulakov, V.I. 2014. Fluorescent analysis of catecholamines and histamine in plant single cells. *International Journal of Biochemistry* 195: 344–351.

Roshchina, V.V., Yashina, A.V., Yashin, V.A., Goltyaev, M.V. 2011b. Fluorescence of Biologically Active compounds in Plant Secretory Cells. In: *Research Methods in Plant Science*, eds. S.S. Narwal, P. Pavlovic and J. John, vol. 2, 3–25, Forestry and Agroforestry. Houston, TX: Studium Press.

Roshchina, V.V., Yashin, V.A., Yashina, A.V., Goltyaev, M.V. 2013. Microscopy for modeling of cell-cell allelopathic interactions. In: *Allelopathy. Current Trends and Future Applications*, eds. Z.A. Cheema, M. Farooq and A. Wahid, 407–427. Berlin, Heidelberg: Springer.

Roshchina, V.V., Yashina, A.V., Yashin, V.A., Prizova, N.K. 2009. Models to study pollen allelopathy. *Allelopathy Journal* 23(1): 3–24.

Rostagno, M.A., Arrigo, M.D., Martínez, J.A. 2010. Combinatory and hyphenated sample preparation for the determination of bioactive compounds in foods. *TrAC Trends in Analytical Chemistry* 29(6): 553–561.

Rumala, C.S., Iadhav, A.N., Smillie, T., Fronczek, F., Khan, R.I.A. 2008. Alkaloids from *Heimia salicifolia*. *Phytochemistry* 69(8): 1756–1762.

Rybalko, K.S. 1978. *Natural Sesquiterpene Lactones*. Moscow: Meditsina, 320 p.

Saltiel, J. 1968. Perdeuteriostilbene. The triplet and singlet paths for stilbene photoisomerization. *Journal of the American Chemical Society* 90(23). doi:10.1021/ja01025a026.

Samadi, S., Khadivzadeh, T., Emami, A., Moosavi, N.S., Tafaghodi, M., Behnam, H.R. 2010. The effect of Hypericum perforatum on the wound healing and scar of cesarean. *Journal of Alternative and Complementary Medicine* 16(1): 113–117. PMID 20064022.

Sandhu, N.S., Kaur, S., Chopra, D. 2010. *Equisetum arvense*: Pharmacology and Phytochemistry. *Asian Journal of Pharmaceutical and Clinical Research* 3(3): 146–150.

Santarpia, J.L., Pan, Y.L., Hill, S.C., et al. 2012. Changes in fluorescence spectra of bioaerosols exposed to ozone in a laboratory reaction chamber to stimulate atmospheric ageing. *Optic Express* 20(28): 29867–29881.

Santhi, A., Kala, U.L., Nedumpara, R.J., Kurian, A., Kurup, M.R.P., Radhakrishnan, P., Nampoori, V.P.N. 2004. Thermal lens technique to evaluate the fluorescence quantum yield of a Schiff base. *Applied Physics B* 79(5): 629–633.

Sarin, R .2003. Enhancement of Opium alkaloids production in callus culture of papaver rhoeas Linn. *Indian Journal of Biotechnology* 2: 271–272.

Sárközi, Á., Janicsák, G., Kursinszk, Kéry, L.Á. 2006a. Alkaloid composition of Chelidonium majus L. studied by different chromatographic techniques. *Chromatographia* 63: 81–86.

Sárközi, Á., Móricz, Á.M., Ott, P.G., Tyihák, E., Kéry, Á. 2006b. Investigation of Chelidonium alkaloids by use of a complex bioautographic system. *Journal of Planar Chromatography* 19(110): 267–272.

Sasaki, K. 2016. Fluorescence: A new trait for flowers. *Single Cell Biology* 5(1): 128–129.

Sasaki, K., Kato, K., Mishima, H., et al. 2014. Generation of fluorescent flowers exhibiting strong fluorescence by combination of fluorescent protein from marine plankton and recent genetic tools in Torenia fournieri Lind. *Plant Biotechnology* 31(4): 309–318.

Scheffer, J.J.C., Looman, A., Baerheim Svendsen, A. 1986. The essential oils of three *Origanum* species grown in Turkey. In: *Progress in Essential Oil Research*, ed. E.-J. Brunke, 151–156. Berlin: Walter de Gruyter and Co.

Schempp, Christoph M., Pelz, K., Wittmer, A., Schöpf, E., Simon, J.C. 1999. Antibacterial activity of hyperforin from St John's wort, against multiresistant Staphylococcus aureus and gram-positive bacteria. *Lancet* 353(9170): 2129. PMID 10382704.

Schnabl, H., Weissenböck, G., Scharf, H. 1986. In vivo-microspectrophotometric characterization of flavonol glycosides in *Vicia faba* guard and epidermal cells. *Journal of Experimental Botany* 37(1): 61–72.

Schulz, H.U., Schürer, M., Bässler, D., Weiser, D. 2005. Investigation of the bioavailability of hypericin, pseudohypericin, hyperforin and the flavonoids quercetin and isorhamnetin following single and multiple oral dosing of a Hypericum extract containing tablet. *Arzneimittel-Forschung* 55(1): 15–22. PMID 15727160.

Sgherri, C., Cosi, E., Navari-Izzo, F. 2003. Phenols and antioxidative status of Raphanus sativus grown in copper excess. *Physiologia Plantarum* 118(1): 21–28.

Shafiee-Hajiabad, M., Hardt, M., Honermeier, B. 2014. Comparative investigation about the trichome morphology of common oregano (*Origanum vulgare* L. subsp. vulgare) and Greek oregano (*Origanum vulgare* L. subsp. hirtum). *Journal of Applied Research on Medicinal and Aromatic Plants*. doi:10.1016/j.jarmap.2014.04.001.

Sharma, R., Gupta, R. 2019. Role of acetylcholine system in allelopathy of plants. In: *Neurotransmitters in Plants: Perspectives and Applications*, eds. A. Ramakrishna and V.V. Roshchina, 243–270. Boca Raton, FL: CRC Press.

Sharma, A.D., Sharma, R. 1999. Anthocyanin-DNA copigmentation complex: Mutual protection against oxidative damage. *Phytochemistry* 52(7): 1313–1318.

Shedoeva, A., Leaveslev, D., Upton, Z., Fan, C. 2019. Wound healing and the use of medicinal plants. *Evidence-Based Complementary and Alternative Medicine* 2019: Article ID 2684108. doi: 10.1155/2019/2684108.

Shikov, A.N., Pozharitskaya, O.N., Makarov, V.G., Wagner, H., Verpoorte, R., Heinrich, M. 2014. Medicinal plants of the Russian Pharmacopoeia; their history and applications. *Journal of Ethnopharmacology* 154(3): 481–536.

Shubina, V.S., Lavrovskaya, V.P., Bezgina, E.N., Pavlik, L.L., Moshkov, D.A. 2011. Cytochemical and Ultrastructural characteristics of BHK-21 cells exposed to dopamine. *Neuroscience and Behavioral Physiology* 41(1): 1–5.

Sikorska, E., Khmelinski, I., Sikorski, M. 2012. Analysis of olive oils by fluorescence spectroscopy: Methods and applications. In *Olive Oil - Constituents, Quality, Health. Properties and Bioconversions*, ed. D. Bosckou, 62–88. London: InTechOpen.

Singab, A.N., Bahgt, D., Alsaved, E, Eldahshan, O.A. 2015. Saponins from Genus *Albizia*: Phytochemical and biological review. *Medicinal and Aromatic Plants* S3. doi: 10.4172/2167-0412.S3-001.

Singh, A. 2018. *Herbal Drugs as Therapeutic Agents*. Boca Raton, FL: CRC Press.

Singh, B., Sharma, R.A. 2020. *Secondary Metabolites of Medicinal Plants, 4 Volume Set: Ethnopharmacological Properties, Biological Activity and Production Strategies*. Hoboken, NJ: Wiley.

Singh, M.K.R., Saku, P., Nagori, K., et al. 2011. Organoleptic properties in-vitro and in-vivo pharmacological activities of *Calendula officinalis* Linn, an over review. *Journal of Chemical and Pharmaceutical Research* 3(4): 655–663.

Singh, O., Khanam, Z., Misra, N., Srivastava, M.K. 2011. Chamomile (*Matricaria Chamomilla* L.): An review. *Pharmacology Reviews* 5(9): 82–95.

Singh, V., Mishra, A.K. 2015. White light emission from vegetable extracts. *Scientific Reports* 5. doi:10.1038/srep11118.

Slaninova, I., Slanina, J., Taborska, E. 2008. Fluorescence properties of quarternary benzo[c]phenanthridine alkaloids and their use as supravital DNA probes. *Chemické Listy* 102: 427–433.

Smirnova, A.V., Timofeyev, K.N., Breygina, M.A., Matveyeva, N.P., Yermakov, I.P. 2012. Antioxidant properties of the pollen exine polymer matrix. *Biophysics (Russia)* 57(2): 258–263.

Sonova, K.V. 2015. Use of Solidago in modern medicine. *Nauchnyj Medicinskij Vestnik* 2(2): 61–67.

Spiridonov, N.A., Konovalov, D.A., Arkhipov, V.V. 2005. Cytotoxicity of some Russian ethnomedicinal plants and plant compounds. *Phytotherapy Research* 19(5): 428–432.

Stanley, R.G., Linskens, H.F. 1974. *Pollen Biology, Biochemistry and Management*. Berlin: Springer-Verlag, 307 p.

Steelink, C., Yeung, M., Caldwell, R.L. 1967. Phenolic constituents of healthy and wound tissues in the giant cactus (*Carnegiea gigantea*). *Phytochemistry* 6(10): 1435–1440.

Stransky, K., Valterova, I. 1999. Release of volatiles during the flowering period of *Hydrosme rivieri* (Araceae). *Phytochemistry* 52: 1387–1390.

Strock, C.F., Schneider, H.M., Galindo-Castañeda, T., et al. 2019. Laser ablation tomography for visualization of root colonization by edaphic organisms. *Journal of Experimental Botany* 70(19): 5327–5342.

Swiedrych, A., Kukuła, K.L., Skirycz, A., Szopa, J. 2004. The catecholamine biosynthesis route in potato is affected by stress. *Plant Physiology and Biochemistry* 42(7–8): 593–600.

Suleymanova, F.S., Nesterova, O.V., Matyushin, A.A. 2017. The historical background and prospects of Canadian goldenrod (*Solidago* canadensis L.) herb medicinal use. Zdorov'e I Obrazovanie inXXI Century. *Health and Education Millennium* 19(4): 142–147.

Süntar, Peşin Ipek, Akkol, Küpeli, Esra, Yılmazer, Alper, M., Yeşilada, E. 2010. Investigations on the in vivo wound healing potential of *Hypericum perforatum* L. *Journal of Ethnopharmacology* 127(2): 468–477. PMID 19833187.

Supratman, U., Fujita, T., Akiyama, K et al. 2001. Anti-tumor promoting activity of bufa-dienolides from Kalanchoe pinnata and K. *daigremontiana x tubiflora. Bioscience, Biotechnology, and Biochemistry* 65(4): 947–949.

Surre, J., Saint-Ruf, C., Collin, V., Orenga, S., Ramjeet, M., Matic, I. 2018. Strong increase in the autofluorescence of cells signals struggle for survival. *Scientific Reports* 8(1). doi:10.1038/s41598-018-30623-2.

Svoboda, K.P., Svoboda, T.G. 2000. *Secretory Structures of Aromatic and Medicinal Plants: A Review and Atlas of Microphotographs.* Knighton, UK: Middle Travely, Beguildy: Microscopix. Publ.

Syrchina, A.I., Gorokhova, V.G., Tyukavkina, N.A., Babkin, V.A., Voronkov, M.G. 1980. Flavonoid glycosides of spore-bearing stems of *Equisetum arvense. Chemistry of Natural Compounds* 16(3): 245–248.

Syrchina, A.I., Voronkov, M.G., Tyukavkina, N.A. 1974. Apigenin-5- glycoside from Equisetum arvense. *Chemistry of Natural Compounds* 5:665–666.

Syrchina, A.I., Voronkov, M.G., Tyukavkina, N.A. 1978. Flavones of *Equisetum arvense. Chemistry of Natural Compounds* 6: 807–808.

Talamond, P., Verdeil, J.L., Conėjėro, G. 2015. Secondary metabolite localization by auto-fluoescence in living plant cells. *Molecules* 20(3): 5024–5037.

Taylor, D.L., Salmon, E.D. 1989. Basic fluorescence microscopy. In: *Methods in Cell Biology: Living Cell in Culture*, eds. J.L. Wang and D.I. Taylor, 207–237. San Diego, NY: Academic Press.

Thiem, B., Goślińska, O. 2002. Antimicrobial activity of *Solidago virgaurea* L. from in vitro cultures. *Fitoterapia* 73(6): 514–516.

Thiengsusuk, A., Boonprasert, K., Bangchang, K.Na. 2019. A systematic review of drug metabolism studies of plants with anticancer properties: Approaches applied and limitations. *European Journal of Drug Metabolism and Pharmacokinetics.* doi:10.1007/s13318-019-00582-8.

Tran, G. 2015. *Russian Comfrey (Symphytum × Uplandicum).* Feedipedia, a programme by INRA, CIRAD, AFZ and FAO. https://www.feedipedia.org/node/92, Last updated on October 5, 2015, 13:16.

Tretyn, A., Kendrick, R.E. 1991. Acetylcholine in plants: Presence, metabolism and mechanism of action. *Botanical Review* 57(1): 33–73.

Tswett, M. 1911. Über Reicherts Fluoreszenzmikroskop und einige damit Angestellte Beobachtungenüber chlorophyll und Cyanophyll. *Berichte der Deutschen Botanischen Gesellschaft* 29: 744.

Tunzi, M.G., Chu, M.Y., Bain, R.C. Jr. 1974. In vivo fluorescence, extracted fluorescence, and chlorophyll concentrations in algal mass measurements. *Water Research* 8(9): 623–635.

Turi, C.E., Axwik, K.E., Murch, S.J. 2014. In vitro conservation, phytochemistry, and medicinal activity of *Artemisia tridentate* Nutt.: Metabolomics as a hypothesis-generating tool for plant tissue culture. *Plant Growth Regulation* 74(3): 239–250.

Turner, G.W., Gershenzon, J., Croteau, R.B. 2000. Distribution of peltate glandular trichomes on developing leaves of peppermint. *Plant Physiology* 124(2): 655–664.

Upyr, T.V., Zaitsev, G.P., Koshovyi, O.M., Komisarenko, A.M. 2015. Phenolic composition of the *Ledum palustre* shoots liquid extract. *Znpsnmapo* 24(5): 49.

Ustyuzhanin, A.A., Konovalov, D.A., Shreter, A.I., Konovalova, O.A., Rybalko, K.S. 1987. The Content of Chamazulenes in *Achillea millefolium* L. in European Part of USSR. *Rastitelnye Resursi (Plant Resources, Russia)* 3: 424–428.

Valette, C., Andary, C., Geiger, J.P., Sarah, J.L., Nicole, M. 1998. Histochemical and cyto-chemical investigations of phenols in roots of banana infected by the burrowing nematode Radopholus similis. *Phytopathology* 88(11): 1141–1148.

van Gijzel, P. 1961. Autofluorescence and age of some fossil pollen and spores. *Koninklijke Nederlandse Akademie van Wetenschappen.* Proceedings Aeries B Physic. Science 64(1): 56–63.

van Gijzel, P. 1967. Autofluorescence of fossil pollen and spores with special reference to age determination and coalification. *Leidse Geologische Mededelingen* 40: 263–317.

van Gijzel, P. 1971. Revie on the ultra-violet-fluorescence microphotometry of fresh and fossill exines and exosporia. In: *Sporopollenin*, eds. J. Brooks, P.R. Grant, M. Muir and van P.R. Gijzel, Proc. Symp. at Geology Department, Imperial college, London, 23–25 September 1970, 659–685. London and New York: Acad. Press.

van Gurp, M., van Ginkel, G., Levine, Y.K. 1989. Fluorescence anisotropy of chlorophyll a and chlorophyll b in castor oil. *Biochimica et Biophysica Acta* 973(3): 405–413.

van Orden, L.S. 1970. Quantitative histochemistry of biogenic amines: A simple microspectrofluorometer. *Biochemical Pharmacol* 19(1): 1105–1106, IN7-IN8, 1107–1117.

Vattikuti, U.M.R., Ciddi, V. 2005. An overreview on *Hypericum perforatum* Linn. *Natural Product Radiance* 4(5): 368–381.

Veit, M., Beckert, C., Höhne, C., Bauer, K., Geiger, H. 1995. Interspecific and intraspecific variation of phenolics in the genus *Equisetum* subgenus *Equisetum*. *Phytochemistry* 38(4): 881–891.

Veit, M., Geiger, H., Wray, V., et al. 1993. Equisetumpyrone, a styrylpyrone glucoside in gametophytes from Equisetum arvense. *Phytochemistry* 32(4): 1029–1032.

Vekshin, N.L. 2011. Binding of Hoechst with nucleic acids using fluorescence spectroscopy. *Journal of Biophysical Chemistry*, 2: 443–447.

Vidot, K., Devaux, M.F., Alvarado, C., et al. 2019. Phenolic distribution in apple epidermal and outer cortex tissue by multispectral deep-UV autofluorescence cryo-imaging. *Plant Science* 28: 51–59.

Vidot, K., Gaillard, C., Rivard, C., Siret, R., Lahaye, M. 2018. Cryo-laser scanning confocal microscopy of diffusible plant compounds. *Plant Methods* 14: 89.

Vihakas, M., Pälijärv, M., Karonen, M., Roininen, H., Salminen, J.P. 2014. Rapid estimation of the oxidative activities of individual phenolics in crude plant extracts. *Phytochemistry* 103: 76–84.

Vogt, T., Pollak, P., Tarlin, N., Taylor, L.P. 1994. Pollination – Or wound-induced kaempferol accumulation in *Petunia* stigmas enhances seed production. *The Plant Cell* 6(1): 11–23.

Waalkes, T.P., Sjoerdama, A., Greveling, C.R., Weissbach, H., Udenfriend, S. 1958. Serotonin, norepinephrine, and related compounds in bananas. *Science* 127(3299): 648–650.

Wagner, H. 1988. Non-steroid, cardio active plant constituents. In: *Economic and Medicinal Plant Research*, eds. H. Wagner, H. Hikino and N.R. Farnsworth, 17–38. London: Academic Press.

Wagner, H., Bladt, S. 1996. *Plant Drug Analysis.* 2nd edition. Dordrecht and Heidelberg: Springer.

Walaas, E., Walaas, O., Haavaldsen, S. 1963. Spectrophotometric and electron-spin resonance studies of complexes of catecholamines with Cu (II) ions and the interaction of ceruloplasmin with catecholamines. *Archives of Biochemistry and Biophysics* 100: 97–100.

Wang, R., Peng, S., Zeng, R., Wien, L., et al. 2007. Herbicidal potential of essential oils of oregano or marjoram (Origanum spp). *Allelopathy Journal* 20(2): 298–305.

Wehling, K., Niester, Ch., Boon, J.J., Willemse, M.T.M., Wiermann, R. 1989. *p*-Coumaric acid - Monomer in the sporopollenin skeleton. *Planta* 179(3): 376–380.

Wei, Q., Acuna, G., Kim, S., et al. 2017. Plasmonics enhanced smartphone fluorescence microscopy. *Scientific Reports* 7(1), article 2124. doi:10.1038/s41598-017-02395-8.

Weissenböck, G.W., Schnabl, H., Scharf, H., Sachs, G. 1987. Secondary phenolic products on the properties of fluorescing compounds in guard and epidermal cells of *Allium cepa* L. *Planta* 171(1): 88–95.

Werle, E., Raub, A. 1948. Über Vorkommen, Bildung und Abbau biogener amine bei Pflanzen unter besonderer Beruck-sichtigung des Histamins. *Biochemische Zeischrift* 318: 538–553.

Wessler, I., Kilbinger, H., Bittinger, F., Kirkpatrick, C.J. 2001. The non-neuronal cholinergic system: The biological role of non-neuronal acetylcholine in plants and humans. *Japanese Journal of Pharmacology* 85: 2–10.

Williams, C.A. 2013. Specialized dietary supplements equine applied and clinical nutrition. In: *Equine Applied and Clinical Nutrition, Health, Welfare and Performance*, eds. R.J. Geor, P.A. Harris and M. Coenen, 351–366. Amsterdam: Elsevier.

Willemse, M.T.M. 1971. Morphological and fluorescence microscopical investigation on sporopollenin formation at *Pinus sylvestris* and *Gasteria verrucosa*. In: *Sporopollenin*. Proceedings of the Symposium at Geology Department., Imperial College, London, 23–25 September 1970. eds. J. Brooks, P.R. Grant, M. Muir and P.R. van Gijzel, 68–91. London and New York: Acad. Press.

Wink, M. 2015. Modes of action of herbal medicines and plant secondary metabolites. *Medicines* 2(3): 251–286.

Wittig, J., Veit, M. 1999. Analyse der Flavonolglycoside von Solidago-Spezies in einem zusammengesetzten Pflanzenextrakt. *Drogenreport* 12: 18–20.

Wojtyniak, K., Szymański, M., Matławska, I. 2013. Leonurus cardiaca L. (motherwort): A review of its phytochemistry and pharmacology. *Phytotherapy Research: PTR* 27(8): 1115–1120. Epub 2012 Oct 8.

Wolfbeis, O.S. 1985. The fluorescence organic natural product. In: *Molecular Luminescence Spectroscopy. Methods and Applications*, ed. S.G. Shulman, 167–370. New York: Chichester Brisbane: Wiley.

Wolken, W.A.M., Tramper, J., van der Werf, M.J. 2003. What can spores do for us? *Trends in Biotechnology* 21(8): 338–345.

Wright, C.W. ed. 2002. *Artemisia*. London and New York: Taylor & Francis. Boca Raton, MA: CRC Press. ISBN 0-415-27212-2642. doi:10.1371/journal.pone.0195642.

Xiu, Z., Zhu, X., Zhang, D. 2003. A new way for chemical degradation of plastic by natural volatile constituents of *Ledum palustre*. *Chinese Science Bulletin* 48(16): 1718.

Yamamoto, K., Momonoki, Y.S. 2019. Plant acetylcholinesterase plays an important role in the response to environmental stimuli. In: *Neurotransmitters in Plants: Perspectives and Applications*, eds. A. Ramakrishna and V.V. Roshchina, 9–33. Boca Raton, FL: CRC Press.

Yeloff, D., Hunt, C. 2005. Fluorescence microscopy of pollen and spores: A tool for investigating environmental change. *Review of Palaeobotany and Palynology* 133(3–4): 203–219.

Yilmaz, S., Ekinci, S., Yilmaz, B. 2014. Determination of bioactive compounds of Equisetum arvense by gas chromatography-mass spectrometry method. *International Journal of Pharmacology* 1(3): 184–188.

Yılmaz, C., Gökmen, V. 2019. Neuroactive compounds in foods: Occurrence, mechanism and potential health effects. *Food Research International*. doi:10.1016/j.foodres.2019.108744.

Yiyu, Xiamen. *Biological Technology Co., Ltd.* Fujian: China Bee Products.

Yuan, F., Chen, M., Leng, B.Y., Wang, B.S. 2013. An efficient autofluorescence method for the screening *Limonium bicolor* mutants for abnormal salt gland density and salt excretion. *South African Journal of Botany* 88: 110–117.

Zelinskii, V.V. 1947. One of possible causes of the internal quenching of fluorescence of complex organic dyes. *Dokl USSR AcadSci* 56(4): 383–385.

Zhang, L., Wang, H.X., Wang, Y.M., Xu, M., Hu, X.Y. 2017. Diurnal effects on Chinese Wild *Ledum palustre* L. essential oil yields and composition. *Journal of Analytical Sciences, Methods and Instrumentation* 7: 47–55.

Zhang, R.H., Liu, Z.K., Yang, D.S., Zhang, X.J., Sun, H.D., Xiao, W.L. 2018. Phytochemistry and pharmacology of the genus *Leonurus*: The herb to benefit the mothers and more. *Phytochemistry* 147: 167–183.

Zhang, Y.,S., Asico, L.D., et al. 2012. Deficient dopamine D2 receptor function causes renal inflammation independently of high blood pressure. *PLoS One* 7(6): 1–11.

Zhang, Y., Cuevas, S., Asico, L.D., et al. 2012. Deficient dopamine D2 receptor function causes renal inflammation independently of high blood pressure. *PLoS One* 7(6): 1–11.

Zhang, X., Zhu, Y., Li, X., Zhang, B., Ji, X., Dai, B. 2016. A simple, fast and low-cost turn-on fluorescence method for dopamine detection using in situ reaction. *Analitica Chimica Acta* 944: 51–56.

Zheng, S.J., Ma, J.F., Matsumoto, H. 1998. Continuous secretion of organic acids is related to aluminum resistance during relatively long-term exposure to aluminum stress. *Physiologia Plantarum* 103(2): 209–214.

Zhou, Y., Hai-Liang Xin, X., Rahman, H.L., Wang, K.,Peng, S.J.C., Hong Zhang, H. 2015. *Portulaca oleracea* L.: A review of Phytochemistry and Pharmacological Effects. *BioMed Research International* 2015, Article ID 925631, 11.

Zia, M., Chaudhary, M.F. 2007. Effect of amino acids and plant growth regulators on artemisinin production in the callus *of Artemisia absinthium. Pakistan Journal of Botany* 39: 799.

Index